D1612074

Geotechnical & Geohydrological Aspects of

WASTE MANAGEMENT

Edited by Dirk J. A. van Zyl, John D. Nelson, Steven R. Abt, Thomas A. Shepherd

Geotechnical Engineering Program, Civil Engineering Department, Colorado State University, Fort Collins, Colorado

Proceedings of the 9th Annual Symposium on Geo-Aspects of Waste Management, February 1 – 6, 1987
Colorado State University
Fort Collins, Colorado

Library of Congress Cataloging-in-Publication Data

Geotechnical and geohydrological aspects of waste
management

Papers from the symposium entitled: Geo-aspects of waste management, the ninth in a series sponsored by the Geotechnical Engineering Program of the Civil Engineering Dept. at Colorado State University.

Includes bibliographies and index.

1. Factory and trade waste—Congresses. 2. Waste disposal in the ground—Congresses. 3. Engineering geology—Congresses. 4. Hydrogeology—Congresses. 5. Sanitary landfills—Congresses. I. Van Zyl, Dirk J. A. II. Colorado State University. Geotechnical Engineering Program.

TD896.G46 1987 628.4′456 86-27331

ISBN 0-87371-101-7

LEWIS PUBLISHERS, INC.
121 South Main Street, P.O. Drawer 519, Chelsea, Michigan 48118

PRINTED IN THE UNITED STATES OF AMERICA

PREFACE

This book was developed from a symposium held at Colorado State University, Fort Collins, Colorado, February 1–6, 1987. This symposium, "Geo-Aspects of Waste Management," was the ninth in a series sponsored by the geotechnical engineering program of the Civil Engineering Department at Colorado State University.

The first five annual Symposia on Uranium Mill Tailings Management, also held at Colorado State University, focused on the design, construction, and operation of uranium tailings impoundments. The sixth and seventh annual symposia were of broader scope, and included low-level and hazardous waste management so as to increase the sphere of interest and provide a forum by which the technology developed for uranium tailings management could be exchanged with the professionals working in low-level and hazardous waste management. The eighth and ninth symposia continued the process of technology transfer but focused more precisely on the geotechnical and geohydrological aspects of waste management—the two engineering areas of prime importance in the design and operation of waste disposal facilities. This is the only annual symposium that focuses specifically on these engineering aspects of waste management and is dedicated to the integration of the profession's understanding of these technical disciplines over the broad spectrum of waste management problems.

The technical chapters included here document the technology development that has taken place and define the current state of the art. It is hoped that this book, as well as the symposium, will enhance the dialogue among research, industry, and regulatory personnel.

Fort Collins, Colorado
February 1987

ACKNOWLEDGMENTS

The organizing committee for the symposium and the associated minicourses—Dirk van Zyl (chairman), Steven R. Abt, John D. Nelson, Thomas A. Shepherd, and Richard E. Wardwell—would like to thank the authors, registrants, keynote speakers, session chairs, and hosts who participated to make the symposium a success.

Dr. Dirk van Zyl received his bachelor's and honors degrees in civil engineering from the University of Pretoria, Pretoria, South Africa, in 1972 and 1974, respectively. His master's and doctoral degrees, with specialization in geotechnical engineering, were conferred by Purdue University in 1976 and 1979.

Dr. van Zyl is an associate professor in the geotechnical engineering program at Colorado State University and is presently program leader. He was on the faculty of the University of Arizona, Tucson, Arizona, for two years before joining the CSU faculty in August 1984. The rest of his experience includes 18 months with the National Institute for Transportation and Road Research on the Council of Scientific and Industrial Research in Pretoria, South Africa. He conducted research on aspects of soil erosion as well as underdrainage of roads. Dr. van Zyl joined the consulting engineering company of Steffen, Robertson and Kirsten in 1974 and has been associated with their offices in Johannesburg, South Africa; Vancouver, Canada; Denver; and Tucson.

Dr. van Zyl has been involved with all stages of tailings disposal planning, design, and construction. He has published a number of papers in this area. He has been involved particularly with in situ testing of tailings impoundments using the piezocone, and contributed to a special study in this for the Bureau of Mines, U.S. Department of the Interior. His present interests are in risk assessment and probabilistic applications in geotechnical engineering.

Dr. van Zyl is a registered professional engineer in seven states, including Colorado. He is an associate member of the American Society of Civil Engineers, where he is secretary of the Safety and Reliability Committee of the Geotechnical Division. He is also associated with the Society of Mining Engineers and the American Society for Testing and Materials, and is editor of "Waste Geotechnics" (part of *Geotechnical News*) as well as co-editor and publisher of *Heap and Dump Leaching,* an international newsletter.

Dr. Steven R. Abt is an associate professor active in the hydraulics and geotechnical engineering programs in the Department of Civil Engineering at Colorado State University. His degrees are a BCE, an MS in water resources, and a PhD in hydraulics from Colorado State University. Dr. Abt has also completed the U.S. Army Engineer Officer basic and advanced courses and is currently enrolled in Command and General Staff College. He is a registered professional engineer in Colorado.

Dr. Abt has 13 years of general engineering experience. After graduation, and one year's service in the U.S. Army Corps of Engineers, he worked as a consulting water resource engineer for two years in Denver, Colorado, performing drainage, flood plain delineation, water rights analysis, and computer modeling studies. In 1977, Dr. Abt joined the civil engineering faculty at Colorado State University as an instructor, serving in a teaching and administrative capacity. Since 1977, Dr. Abt has become active in research, with interests in hydraulic models, river mechanics, flow measurement, erosion and sedimentation, and tailings management. He has published over 40 journal articles and reports.

Dr. Abt has miscellaneous experience serving as a contract facility engineer and contract officer/technical representative for the Rocky Mountain Station, U.S. Forest Service; embankment inspector for the Nuclear Regulatory Commission; and hydrologic consultant to the state of New Mexico. He has also served as a coordinator, organizer, and administrator of numerous symposia, conferences, and continuing education events. He has earned the Haliburton Engineering Young Faculty Award, the Dow Chemical Outstanding Young Faculty Award, and the Stone and Webster Engineering Corporation Fellowship Award.

Dr. John D. Nelson received his bachelor's, master's, and PhD degrees in civil engineering from Illinois Institute of Technology in 1960, 1962, and 1967. He is a registered professional engineer in four states, and specializes in geotechnical engineering.

Dr. Nelson began his engineering career at the IIT Research Institute in Chicago, Illinois, from 1962 to 1968, as a research engineer responsible for directing research in lunar soils and soil dynamics. From 1968 to 1973 he served on the faculty at the Asian Institute of Technology, a graduate school of engineering in Bangkok, Thailand. In that position he taught graduate courses and conducted research on soil dynamics, strength, and general properties of soft and overconsolidated clays. He also had extensive experience in field instrumentation and soft clays.

From 1973 to 1986 he was program leader of the geotechnical engineering program at Colorado State University, and is now head of the Department of Civil Engineering. At CSU, Dr. Nelson's interests have focused around management of mill tailings, mechanics of expansive and unsaturated soils, and creep failure of overconsolidated clays. He has had extensive experience providing technical assistance to the U.S. Nuclear Regulatory Commission with regard to long-term stability of uranium mill tailings impoundments. He has also conducted some research into liquefaction of soils.

Dr. Nelson has been involved in a number of consulting projects, primarily in areas relating to design of earthfill and tailings dams. He served as a chief technical reviewer for the U.S. NRC on the Church Rock tailings dam failure and has served as design reviewer on a number of other dams. He has also served as a consultant on a number of projects dealing with slope stability, foundation engineering, and blast effects on structures.

Dr. Nelson is active in a number of professional societies. He is a past chairman of the American Society of Civil Engineers Committee on Embankment Dams and Slopes and is currently chairman of the Political Action Committee for the Professional Engineers of Colorado. He is also associated with the Society of Mining Engineers, the National Society of Professional Engineers, and the International Society for Soil Mechanics and Foundation Engineering.

Dr. Thomas A. Shepherd received his bachelor's degree in chemistry from Duke University in 1961. He received an MBA in finance from Indiana University in 1963. He continued his studies at Colorado State University, where he received a MS in resource planning in 1974 and a PhD in the same area in 1979.

Before entering Colorado State University to pursue his interests in environmental engineering issues, he worked for 10 years in various management and technical capacities for a chemical manufacturing company in Cincinnati, Ohio.

Dr. Shepherd is a corporate consultant to, and part owner of, the engineering consulting firm of Water, Waste & Land. He is also a part-time assistant professor in the Department of Civil Engineering of Colorado State University. He teaches graduate level civil engineering courses in the chemical and physical aspects of soil, emphasizing the relationship to waste disposal.

Dr. Shepherd has worked with the mining industry for the last eight years assisting that industry to design, license, operate and reclaim tailings disposal facilities. The uranium mining industry has been his principal client, and his efforts for them have focused on the investigation of groundwater impacts and contaminant migration, and the development of operational and reclamation designs that satisfy regulatory requirements and at the same time provide economic and technically effective solutions for tailings disposal.

CONTRIBUTORS

Steven R. Abt, Civil Engineering Department, Colorado State University, Fort Collins, CO 80523

Jennifer L. Askey, FMC Corporation, Minerals Division, P.O. Box 872, Green River, WY 82935

James M. Beck, ERT, A Resource Engineering Company, 1716 Heath Parkway, Fort Collins, CO 80524

William Black, Westbay Instruments Ltd., 507 E. Third Street, North Vancouver, British Columbia, Canada V7L lG4

W. V. Bluck, CH2M Hill, P.O. Box 22508, Denver, CO 80222

Jack A. Caldwell, Jacobs Engineering Group, Inc., 5301 Central Avenue N.E., Suite 1700, Albuquerque, NM 87108

Hsien Chen, International Technology Corporation, 17500 Red Hill Avenue, Suite 100, Irvine, CA 92714

William E. Cobb, CH2M Hill, P.O. Box 22508, Denver, CO 80222

David E. Daniel, Civil Engineering Department, The University of Texas at Austin, Austin, TX 78712

Lyle Davis, Water, Waste & Land, Inc., Creekside Two Building, 2629 Redwing Road, Suite 200, Fort Collins, CO 80526

Lewis T. Donofrio, Jr., Department of Civil Engineering, Drexel University, 32nd Street and Chestnut Street, Philadelphia, PA 19104

Wesley S. Ethiyajeevakaruna, Civil Engineering Department, 2304B Engineering Building, University of Wisconsin-Madison, Madison, WI 53706

Frank Gontowski, Department of Civil Engineering, Drexel University, 32nd Street and Chestnut Street, Philadelphia, PA 19104

Kingsley Harrop-Williams, The BDM Corporation, 7915 Jones Branch Drive, McLean, VA 22102–3396

Bruce R. Hensel, Illinois State Geological Survey, 615 East Peabody Drive, Champaign, IL 61820

Beverly L. Herzog, Illinois State Geological Survey, 615 East Peabody Drive, Champaign, IL 61820

Tim Holbrook, ERT, A Resource Engineering Company, 1716 Heath Parkway, Fort Collins, CO 80524

Jon W. Hughes, AWARE Incorporated, 80 Airport Road, West Milford, NJ 07480

Ian P. G. Hutchison, Steffen, Robertson & Kirsten, 3232 S. Vance, Suite 210, Lakewood, CO 80227

R. Janardhanam, Department of Civil Engineering, The University of North Carolina at Charlotte, Charlotte, NC 28223

Jey K. Jeyapalan, Wisconsin Hazardous Waste Management Center, 2304B Engineering Building, University of Wisconsin-Madison, Madison, WI 53706

Kathryn Johnson, GECR, Inc., P.0. Box 725, Rapid City, SD 57709

Thomas M. Johnson, Levine-Fricke Consulting Engineers and Hydrogeologists, 629 Oakland Avenue, Oakland, CA 94611

Ned Larson, Jacobs Engineering Group, Inc., 5301 Central Avenue N.E., Suite 1700, Albuquerque, NM 87108

Joseph P. Martin, Department of Civil Engineering, Drexel University, 32nd Street and Chestnut Street, Philadelphia, PA 19104

Edward Mehnert, Groundwater Section, Hazardous Waste Research and Information Center, Illinois State Geological Survey, 615 East Peabody Drive, Champaign, IL 61820

D. Douglas Miller, Rogers & Associates Engineering Corp., P.O. Box 330, Salt Lake City, UT 84110–0330

Jerry R. Miller, Illinois State Geological Survey, 615 East Peabody Drive, Champaign, IL 61820

Louis L. Miller, Water, Waste & Land, Inc., Creekside Two Building, 2629 Redwing Road, Suite 200, Fort Collins, CO 80526

Michael Monteleone, AWARE Incorporated, 80 Airport Road, West Milford, NJ 07480

M. Granger Morgan, Department of Engineering and Public Policy, Carnegie-Mellon University, Pittsburgh, PA 15213

Wesley K. Nash, Jr., FMC Wyoming Corporation, P.0. Box 872, Green River, WY 82935

Kirk K. Nielson, Rogers & Associates Engineering Corp., P.0. Box 330, Salt Lake City, UT 84110–0330

S. Jay Olshansky, National Opinion Research Center, University of Chicago, Chicago, IL 60680

A. Perumal, Civil Engineering Department, Johnson C. Smith University, Charlotte, NC 28216

Ron Rager, Sergent, Hauskins & Beckwith, 4700 Lincoln Road N.E., Albuquerque, NM 87109

Eric Rehtlane, Westbay Instruments Ltd., 507 E. Third Street, North Vancouver, British Columbia, Canada V7L lG4

Larry D. Rickertsen, Rogers & Associates, c/o Weston, 955 L'Enfant Plaza S.W., Eighth Floor, Washington, DC 20024

John C. Robins, Lippincott Engineering Associates, 501 Burlington Avenue, Delanco, NJ 08075

Vern C. Rogers, Rogers & Associates Engineering Corp., P.0. Box 330, Salt Lake City, UT 84110-0330

Raj K. Singhal, CANMET, P.O. Bag 1280, Devon, Alberta, Canada T0C lE0

W. P. Staub, Energy Division, Oak Ridge National Laboratory, Oak Ridge, TN 37831

T. Vladut, RETOM 1985 Research & Development Ltd., 200 2749 39th Avenue N.E., Calgary, Alberta, Canada TlY 4T8

John F. Wallace, Dames & Moore, 250 East Broadway, Suite 200, Salt Lake City, UT 84111-2480

Larry W. Well, CH2M Hill, P.0. Box 4400, Reston, VA 22090

R. Gary Williams, Hazards Assessment Laboratory, Department of Sociology, Colorado State University, Fort Collins, CO 80523

Joe Winch, Civil Engineering Department, 2304B Engineering Building, University of Wisconsin-Madison, Madison, WI 53706

Leonard O. Yamamoto, International Technology Corporation, 27500 Red Hill Avenue, Suite 100, Irvine, CA 92714

CONTENTS

Geotechnical & Geohydrological Aspects of

WASTE MANAGEMENT

GEOMEMBRANES AND SURFACE IMPOUNDMENTS--DESIGN AND CONSTRUCTION ASPECTS

Larry W. Well,
CH2M HILL, Reston, Virginia

INTRODUCTION

Impoundment facilities are as old as mankind itself. Natural rivers, lakes, ponds, and pools conveyed, collected, or stored water and thereby provided a nucleus for primitive communities. Dry holes, pits, ravines, and canyons were convenient locations for disposal of unneeded, unwanted materials. Throughout the thousands and thousands of years of growth and evolution of mankind, fewer natural impoundments such as these were left available near the desirable living locations. As the frontiers of population moved outward, more and more settled areas lacked the natural voids to serve as impoundments. Enter the "evolutionary" engineer to fill that void by creating a void and then filling it. Primitive investigations of natural facilities that worked were undoubtedly carried out by the evolutionary engineer to learn what made them work. As a result of this early version of the scientific method, the functional attributes of those facilities were determined and incorporated into the primitive designs of the evolutionary engineer. The evolutionary contractor undoubtedly did the work.

Until about 20 years ago, we were basically following the precepts and practices of the evolutionary engineer in using finer and finer soils, and eventually asphaltic and portland cement, to construct impoundments for liquid materials. Pits, holes, ravines, valleys, etc., were still

Geotechnical and Geohydrological Aspects of Waste Management, D. J. A. van Zyl et al., Eds., © 1987 Lewis Publishers, Inc., Chelsea, Michigan—Printed in USA.

being used for placement of solid waste materials, as had been done for thousands of years.

But what happened to change all of this? Lots! For example:

- o Concern for the environment, i.e., protection of air, water, and land. U.S. EPA, RCRA, CERCLA, etc.
- o Development of products previously unavailable. Geomembranes, geotextiles, geogrids, and geosynthetics
- o Competition to improve classical approaches to compete with new products by further development of soil admixtures such as bentonite, cement, and asphalt

In the last 20 years, these changes have led the designer in new directions. Comprehensive, complicated programs to protect the environment have been developed, and laws have been written that govern design. A host of new products have changed our way of doing things; new products and new design possibilities using these products open up a whole range of new approaches to designs.

New industries have been formed to produce and market these new materials. Engineering schools are presenting courses relating to materials and construction methods that barely existed 20 years ago. The purchasers of these engineering services have become very informed clients as they learn what they must do to address their needs for environmental protection and control.

All these changes present new problems and challenges to the designer. Technically sophisticated clients, materials, construction methods, and engineers are all factors in modern impoundment facility design. This new technology is now routinely applied to surface impoundments. Examples of some of these impoundments are:

- o Water reservoirs with or without floating covers for raw and potable water
- o Waste sludge storage and/or solidification basins
- o Solid waste burial and control facilities
- o Wastewater storage and/or treatment systems

Every facility used for surface impoundment has similarities as well as differences. Each facility must be evaluated and designed separately to meet the goals and achieve the function, performance, and safety expected by all parties. The design and construction considerations may be similar, but the functions of the facilities may vary widely.

The most important, common, and demanding similarity for each facility is actual containment or "water-tightness" for the liquid or liquid phase of extracts from solids. Another similarity is that most facilities are built at or near natural ground level. Some may make effective use of natural depressions. They are also constructed primarily using solid and earth materials as the predominant building material. Many facilities incorporate synthetic materials to improve their function or to serve in place of natural barriers or drainage.

The differences between impoundments are primarily related to the materials contained and the functions required of the facility. The materials may be liquid or solid, and they may be useful, objectionable, or hazardous. The function may be (1) storage for use or treatment or (2) storage for short- or long-term disposal.

The job of the designer is made more challenging by the fact that traditional engineering approaches must be used in conjuction with new materials on behalf of an informed client who is charged with meeting strict performance criteria. Since all these activities involve people making choices and decisions, the human element surfaces. We all want to know what to expect.

EXPECTATIONS

The expectations for the facilities are as varied as the facilities themselves. Everyone involved has expectations; some are preconceived, some are implied, and some are justified. Unfortunately, not all parties are preconceiving, implying, or justifying the same expectations. A complex matrix of interactions and misunderstandings is commonly encountered during the concept, design, and permitting process of a project, and this matrix forms the crux of our problems as designers or constructors of impoundment facilities. Our most important role as designers should be to clarify preconceived ideas and expectations, to refrain from making implied promises, and to define and assist in achieving justifiable expectations.

For example, the owner's preconceived expectations are for an economical, functional, low-maintenance, leakproof, state-of-the-art facility, meeting all the regulatory requirements, to be completed on time and within budget. The owner's implied expectations are often derived from hearing and reading about "the best" of similar projects and from "sales pitches" wherein performance is alluded to but not promised or guaranteed. The preconception based on this review of information often leads owners to believe the project will be a simple one. The owners' justified expectations (and the ones they have a right to receive) relate to a well engineered, designed, and constructed facility for the budget they have made available.

It is not at all unusual for the owner to be surprised at the total cost for design and construction. The prudent engineer will involve the owner during concept and design activities in order to provide a means of participation and an understanding of the cost development. This involvement may be one of the single most effective means of achieving the owner's expectations. There are, of course, limits as to what can be achieved for the budget amount. At the onset, this point must be made perfectly clear...the budget controls.

The regulatory agencies also have their sets of preconceived, implied, and justified expectations. Preconceived: Cost no object for either the design or construction. Implied: State-of-the-art design that is leakproof and water-tight. Justified: Compliance and performance as mandated by laws, standards, or regulations.

The material manufacturers, suppliers, and installers have their sets of expectations as well. The manufacturers, for example, expect their materials to be accepted at face value. The installers plan to install the materials in the shortest possible time, and often without considering other project activities that may interfere. This applies to suppliers of natural as well as synthetic materials.

The general contractors often fail to plan for the effort and time required to construct an impoundment. They may be unfamiliar with the special materials and special methods of installation they required. Project plans and specifications should present as clear a picture of what to expect as possible.

The designers are also operating under their own set of preconceived, implied, and justified expectations. They must juggle the expectations of all the parties as they go through the concept planning, design, and plans and specifications preparation phases. It is incumbent upon the designers, who are ultimately held or named as the "responsible party," to do the very best job they can to clarify all expectations for all parties at the earliest possible time.

CONCERNS OF DESIGNERS

Designers' concerns are fueled and driven by a matrix of expectations from all parties, as well as their own. Impoundment designs are usually a team effort, and even on the same team some differences in philosophy of approach emerge. When all these expectations conflict, designers (as the responsible parties or experts) have to be prepared to resolve the issues and defend their actions and recommendations. On any given project, they have to successfully deal with the following concerns, which are not presented in order of importance since they are all important:

- Gathering sufficiently detailed information on the site, process, or products related to the containment
- Developing a design budget that will allow the best application of the technology to the unique site, process, or product
- Understanding regulatory constraints and ways of achieving compliance with federal, state, and local requirements
- Explaining to the owner the potential and the limitations of each site, process, or product to be used
- Assigning senior review responsibilities to qualified persons at the beginning, intermediate, and final stages of the project
- Handling design details that were changed or compromised during construction
- Ensuring constructability of the design by providing adequate plans and specifications
- Developing an effective quality control/ quality assurance program

- o Providing qualified, experienced observers during construction
- o Accommodating changing conditions within the site, process, or products
- o Responding to changing regulatory requirements

All of these concerns must be addressed in such a way as to show that we are meeting the standards of the design profession. We are being directed by regulations to make construction materials perform to the limit of and beyond their known capabilities. Regulations in the future may be changed by long-term performance evaluations of the new materials and methods. Many of the materials in common use today have relatively limited performance histories. Each group of materials, whether soils, soil admixtures, clays, geotextiles, geogrids, or flexible membrane linings, requires strict control and inspection during fabrication and installation. The documentation of all design and construction activities is vital to providing a clear record of design decisions and construction of the surface impoundment. The level of effort to achieve the desired results is always underestimated. This is why clarification of expectations described earlier is so important.

CONCLUSION

When we resolve our concerns as designers, we will have gone a long way toward mollifying the concerns of the other parties over unfulfilled expectations. With the development of the modern synthetic materials and their applications, we ourselves may eventually be remembered as the "evolutionary engineers" of this age.

The record will show that the successful completion of impoundment facilities that meet all of the performance expectations is directly proportional to the effort required to develop the design and the diligent, full-time inspection of the installation. Whenever we do less that that, we can expect to have problems. Let us do our best.

WDR205/029

NATIONAL RESEARCH COUNCIL PANEL REPORT--SCIENTIFIC BASIS FOR RISK ASSESSMENT AND MANAGEMENT OF URANIUM MILL TAILINGS

M. Granger Morgan, Panel Chairman
Carnegie Mellon University, Pittsburgh, Pennsylvania

EXECUTIVE SUMMARY

Uranium mill tailings are the finely ground sand-like material that is left after uranium is extracted from ore. As of early 1983, approximately 200 million metric tons of uranium mill tailings covering 1300 ha at 51 sites had accumulated in the United States. Most of these sites are in the arid Southwest.

Uranium mill tailings present health and environmental concerns because of the residual radioactivity that they contain and because of a variety of other potential pollutants, such as chlorides, sulfates, and heavy metals. The milling process makes the radioactive and nonradioactive materials contained in the ore when it was mined much more mobile and also adds several potential contaminants to the tailings material during the milling process. Exposure routes of concern are release of the gas ^{222}Rn, airborne dust, and surface and groundwater contamination. In addition, the use of tailings as construction material or fill can lead to dangerously high levels of radon in associated buildings.

At the request of the Department of Energy, a National Research Council study panel, convened by the Board on Radioactive Waste Management, has examined the scientific basis for risk assessment and management of uranium mill tailings and issued this final report containing a number of recommendations. Chapter 1 provides a brief introduction to the problem. Chapter 2 examines the processes of uranium extraction and the mechanisms by which radionuclides and toxic chemicals contained in the ore can enter the environment. Chapter 3 is largely devoted to a review of the evidence on health risks associated with radon and its decay products. Chapter 4 provides a consideration of conventional and possible new technical alternatives for tailings management. Finally, Chapter 5 explores a number of issues of comparative risk, provides a brief history of uranium mill tailings regulation, and

Geotechnical and Geohydrological Aspects of Waste Management, D. J. A. van Zyl et al., Eds., © 1987 Lewis Publishers, Inc., Chelsea, Michigan – Printed in USA.

concludes with a discussion of choices that must be made in mill tailing risk management.

Several major themes recur throughout the report:

- To ensure the effectiveness of control measures over the long term, it will be necessary to maintain a low level but ongoing program of surveillance of uranium mill tailings piles with provisions for any necessary corrective actions.
- Because of the variability of physical, chemical, radiological, and demographic factors at each site, risk-management strategies must be site specific.
- The health risks posed by exposure to radon from uranium mill tailings piles are trivial for the average U.S. citizen, range from small to modest for most persons who live close to piles, but in special circumstances could be significant for a few individuals who live in close proximity to certain uncontrolled piles.

The principal recommendations of the panel are set forth below.

A major focus of regulation and risk management is control of radon emitted from tailings piles. The relationship between the concentration of radionuclides in a tailings pile and the flux of radon that leaves the pile is complex. The relationship between the flux of radon leaving a pile and the concentration of radon in air at points surrounding a pile is also complex. There is considerable pile-to-pile variability in both of these relationships.

> Recommendation: In developing or revising a risk-management strategy to control the potential risk of radon released from uranium mill tailings piles, the panel recommends a site-specific approach because of the great variability in radon flux at different piles.

While the basic factors that give rise to the complexity and variablity in the relationships between the concentration of radionuclides in a pile, the flux of radon leaving a pile, and the concentration of radon in the air around the pile are understood, models that attempt to relate pile concentrations to radon flux or radon flux to air concentrations give rise to estimates that may be in error by as much as several orders of magnitude.

> Recommendation: In developing or revising a radon risk-management strategy for tailings piles, undue reliance should not be placed on general models of radon emission and dispersion.

Despite the existence of simple, inexpensive, and reliable passive integrating radon monitors, only a modest and incomplete set of measurements of radon in the vicinity of controlled and uncontrolled piles has been collected.

The existing set of field measurements around uncovered tailings piles show that, at some piles, average radon concentrations fall to background levels at a distance of less than a kilometer, and at all measured piles they fall to background within a few kilometers.

> Recommendation: There is a clear need for a systematic program of measurements of radon concentrations around piles before and after the implementation of radon control strategies. These measurements should include an adequate determination of background levels and the generation of concentration isopleths out to background.

Brief periods of time (e.g., hours to days) spent outdoors near uranium mill tailings piles pose no significant lung-cancer risk. Persons of average lifestyle living in close proximity to uncontrolled uranium mill tailings piles may, depending on site-specific circumstances, experience a significant increase in total lifetime radon lung-cancer risk. If a person were to live right at the edge of a few piles that involve particularly unfavorable exposure conditions, their lung-cancer risk could be substantially higher than the average U.S. population lung-cancer risk. However, persons living at distances greater than a kilometer from most uncontrolled uranium mill tailings piles, and perhaps somewhat closer to some piles, will experience no significant increase in lifetime radon lung cancer risk due to exposure to radon from the pile.

> Recommendation: People who are not occupationally involved in uranium milling should not live or spend a large fraction of their time in close proximity to uncontrolled uranium mill tailings piles. If for political, economic, or other reasons it is not feasible to preclude people from living or spending a significant fraction of their time in close proximity to an uncontrolled mill tailings pile, then by almost any risk management decision criterion, steps should be taken to control radon emissions from that pile.

Radon flux from a cover of less than a few meters of earth cannot now be predicted, with precision, either immediately after construction of the cover or in the long term when various physical changes, such as change in water content, have taken place in the cover.

> Recommendation: The design of covers for the purpose of limiting radon release should be validated on a site-specific basis by measurements of radon concentrations in the vicinity of the pile. Such measurements could lead to a reduction in the minimum thickness of earth cover needed to limit radon release to acceptable levels.

Long-term irregular settlement of tailings piles may cause mechanical disruption of any type of cover (earth, asphalt, stabilized tailings, plastic sheeting, or some combination) with consequent changes in rate of radon emission and rainfall infiltration. Chemical deterioration of cover materials may also change their effectiveness over time.

> Recommendation: Periodic inspection and maintenance are essential to ensure that covers of any type will continue to function effectively over the long term.

Because it contains radionuclides, airborne dust from uranium mill tailings piles presents a small, but real, health risk to persons who might inhale such dust. In addition, windblown dust may also present a local health risk from increased external gamma radiation around some piles. Based on available data, the increment of radionuclides from pile dust is indistinguishable from natural background at distances commonly less than a kilometer and never more than a few kilometers; however, the data, particularly data on background levels, are sparse. Strategies to reduce radon emanation appear likely to control the problem of windblown dust.

> Recommendation: Piles should be controlled in a manner that prevents pile material from being blown away as dust.

Even over extended periods of time, the spatial extent of groundwater contamination from most uranium mill tailings is likely to be limited to dimensions of between several hundred and several thousand meters. However, uncontrolled uranium mill tailings piles hold the potential to produce significant local contamination of groundwaters and surface waters. The pattern and nature of this contamination display considerable complexity and a large amount of pile-to-pile variability.

> Recommendation: The specific strategies adopted to control groundwater and surface-water contamination from uranium mill tailings piles should be selected and implemented on a site-specific basis.

Liners that are installed under tailings piles are almost certain in time to fail because of construction flaws, mechanical movements caused by the weight of the tailings, chemical changes associated with the leachate, and other factors.

> Recommendation: If liners are used, and local groundwater protection is deemed essential, they should be backed up by a drainage system for collecting leachate that may leak through the liner and by a system to monitor the effectiveness of the liner/drainage system. A contingency plan should be developed for treating any contamination that does enter the groundwater.

In the case of existing piles, leachate from the tailings may reach groundwater.

> Recommendation: Groundwater monitoring should be undertaken to evaluate the need for remedial action.

When surface-water and groundwater problems arise from uranium mill tailings piles, nonradioactive contaminants may pose the principal water-quality problem. Such water contamination may have many features in common with contamination from mill tailings produced in recovery of some other metals and with that emanating from solid-waste disposal sites.

> Recommendation: It is recommended that surface-water and groundwater contamination releated to uranium mill tailings not be considered as a separate water-quality problem but rather in the context of the broader issue of water contamination emanating from man-made accumulations in general.

While in special circumstances, tailings may pose health risks through a variety of secondary pathways, such as the food chain, in general, it does not appear that such risks are significant. However, if tailings are misused by man as construction material and backfill for occupied buildings, they can pose a serious health risk.

> Recommendation: Successful protection against inappropriate use of tailing materials by man requires a low-level ongoing program of societal recordkeeping (e.g., maps), land-use control, and occasional on-site inspection.

Based on a consideration of the geological evidence, the stability of tailings accumulations in surface or near-surface piles cannot be assured over time periods of more than thousands of years (and in some cases hundreds of years) without the option of active human intervention.

> Recommendation: Plans for the management of mill tailings piles should recognize and explicitly incorporate plans for a continued low-level program of active monitoring and the option for active human intervention should it become necessary.

Protection of some piles against catastrophic floods will not be possible even with heroic measures. However, the risk posed by piles under such circumstances is small and inconsequential compared with other impacts of such a catastrophic event, because tailings would almost certainly be mixed with and diluted in a much larger volume of sediments.

> Recommendation: While protection of piles against local periodic floods is feasible and appropriate, protection against large, truly catastrophic floods is neither feasible nor warranted from the perspective of the risk presented.

Solidification of tailings by any of several techniques may be a useful option in selected special circumstances, but because of high costs, uncertainties as to long-term stability, and problems presented for possible future reprocessing, it is not an attractive general solution.

Modifications and alternatives to traditional processing technologies hold potential for possible reprocessing of old ore and improving the future milling of new ore.

Both in reprocessing existing tailings and in applying various new or modified processes in new extraction, most of the thorium and much of the radium could be removed. These processes would produce relatively small volumes of thorium- and radium-bearing concentrates, the disposal of which might pose significant technical and political problems, although it is conceivable that a market for these materials could develop in the future. New process technologies may also be more efficient than those currently in use. For these reasons, reprocessing of tailings and the development of new process technologies should be evaluated both as a strategy for hazard management and in economic terms.

> Recommendation: A program of research on the process technologies that might be applied in reprocessing should be considered.

> Recommendation: An expanded program of research on technologies that might be used to modify or replace existing processes in order to reduce or eliminate radiological hazards in tailings management should be undertaken.

Viewed in the perspective of the wide variety of risks that face U.S. society, simple order-of-magnitude arguments and comparisons suggest that the health risks posed by exposure to radon from uranium mill tailings piles are trivial for the average U.S. citizen and range from small to modest for most persons who live close to uncontrolled piles. However, if persons were to live right at the edge of a few uncontrolled piles that involve particularly unfavorable exposure conditions, their risk could be significant.*

The controls now being implemented for inactive piles will be more than sufficient to manage the risks that they may pose for the next few decades. However, further risk analysis is needed to understand how best to manage the risks of piles now classified as inactive and to understand the requirement for ongoing observation and corrective intervention on piles now being controlled. The panel makes a number of recommendations for improving future risk assessments, observing that those undertaken to date in the United States have not made adequate use of modern techniques for the characterization and treatment of the substantial scientific uncertainties that are involved.

The panel's findings carry a number of implications for the ongoing regulation of risk from uranium mill tailings.

---

*For this case, the incremental lung-cancer risk per 70-year lifetime is of the order of 0.06, or on an annual basis roughly 8×10^{-4}. For purposes of comparison, this is roughly two-thirds of the individual lung-cancer risk of smoking, or three times the per capita risk of motor vehicle accidents.

Recommendation: If regulation is to be based on the science of the problem, generic models must be validated for actual conditions.

Recommendation: In future regulatory reviews of uranium mill tailing risk management, the U.S. Environmental Protection Agency and the U.S. Nuclear Regulatory Commission should

- Be explicit about the decision criteria that they chose to apply;
- Address the need for ongoing observation and corrective intervention that the science of the problem indicates will be necessary;
- Work to separate and identify the "judgment calls" explicitly, to identify their implications clearly, and to justify the choice compared with other possible alternatives.

Recommendation: The U.S. Environmental Protection Agency should strive to achieve greater internal consistency in its approach to this problem and greater consistency between the way in which it deals with risks from tailings and the way in which it has dealt with other similar risks.

HYDRAULIC CONDUCTIVITY TESTS FOR CLAY LINERS

David E. Daniel,
The University of Texas at Austin

INTRODUCTION

The single most important characteristic of a clay liner is its hydraulic conductivity (permeability). If a clay liner is to function properly, it must have a low hydraulic conductivity. Many regulatory agencies require the hydraulic conductivity of clay liners be less than some specified value (usually less than 1×10^{-7} cm/s). There is evidence that if the hydraulic conductivity can be made to be less than about 1×10^{-8} cm/s, release of contaminants through the clay liner will be primarily via molecular diffusion rather than advective transport [11, 18].

In recent years, the question of how one should go about determining the hydraulic conductivity of clay liners has been the subject of much discussion. The main issues have been:

1. Are laboratory hydraulic conductivity tests adequate, or are field (in situ) tests necessary?

2. If laboratory tests are used, which type of test should be performed and how should the test be conducted?

3. If in-situ tests are used, which type of test should be performed and how should the test be conducted?

The first question listed above is a critical one. With compacted clay liners, several investigators have found that laboratory hydraulic conductivity tests have yielded hydraulic conductivities that are much too low [6, 9, 17, and others]. The author knows of only one instance in the open literature in which laboratory hydraulic conductivity tests are known to have yielded correct values for a compacted clay liner [16], and this case involved extraordinary care in construction of a homogeneous liner built of clayey sand. If a clay liner is constructed such that it is homogeneous and free of hydraulic defects (cracks, fissures, and inter-clod voids), then there is every reason to believe that laboratory and field hydraulic conductivity tests will yield the same results. However, how can one be assured that the liner is sufficiently homogeneous and free of defects to make such an assumption unless one performs both laboratory and in-situ tests?

Geotechnical and Geohydrological Aspects of Waste Management, D. J. A. van Zyl et al., Eds.,

For this reason, the author believes that in-situ hydraulic conductivity tests should always be used in some form to study the hydraulic conductivity of a compacted clay liner. Laboratory hydraulic conductivity tests are still needed for construction quality assurance and to study soil-waste interactions but, in the author's opinion, should not be used as the sole basis for predicting the overall hydraulic conductivity of a compacted clay liner.

Naturally-occurring clay liners are often used in waste management projects. The critical question about such liners is whether they are cracked or contain other types of hydraulic defects. Daniel, Trautwein, and McMurtry [7], Griffin et al. [12], and Keller, van der Kamp, and Cherry [14] provide examples of natural clay liners that were thought to be intact and of low hydraulic conductivity but which were much more permeable than would be suggested on the basis of laboratory hydraulic conductivity tests. On the other hand, Goodall and Quigley [11] found a naturally-occurring stratum of clay near Sarnia, Ontario, Canada, to be relatively impermeable in the field. Thus, as with compacted clay, laboratory hydraulic conductivity tests will yield proper values of in-situ hydraulic conductivity only if the formation under consideration is homogeneous and free of hydraulic defects such as cracks, fissures, and sand or silt seams. Laboratory hydraulic conductivity tests are the only practical means for studying the effects of chemicals or waste liquids upon the hydraulic conductivity of soils.

From this background discussion, it should be clear that a rigorous evaluation of the hydraulic conductivity of a clay liner (natural or man-made) is a multi-faceted problem requiring an understanding of stratigraphy and micro- and macro-morphology of the soil. While the hydraulic conductivity of the soil is the critical parameter, simply performing hydraulic conductivity tests without examining all the variables that can affect hydraulic conductivity is ill advised. Hydraulic conductivity tests should be viewed as one of the many tools available to the engineer or scientist to assist in evaluating the overall hydraulic conductivity of a clay liner. The use of experience and judgment should always be an important component in the evaluation of the overall hydraulic conductivity of a clay liner. After all, no test (even in-situ test) can possibly be performed to permeate all of the clay liner material. Tests only provide a measure of the average hydraulic conductivity of a limited volume of soil; because of this, hydraulic conductivity tests only provide part of the answer to the question: What is the overall hydraulic conductivity of a clay liner?

In the remainder of this paper, the procedures for performing hydraulic conductivity tests will be discussed. Most of the attention will be focused on in-situ tests because less has been written about how those tests should be performed compared to laboratory tests. Further, due to space limitations, emphasis will be given to compacted clay liners rather than natural liners.

LABORATORY HYDRAULIC CONDUCTIVITY TESTS

Types of Permeameters

Permeameters can be grouped as follows:

1. Fixed-Wall Permeameters
   - A. Compaction-Mold Permeameter
   - B. Consolidation-Cell Permeameter
   - C. Sample-Tube Permeameter
2. Flexible-Wall Permeameter

Fixed-wall permeameters generally enjoy the advantages of simplicity and low cost, but suffer from potential problems in back-pressure saturating the soil, confirming saturated conditions, and sidewall leakage. The potential problem in back-pressurization of a fixed-wall permeameter is that the back pressure may expand the rigid confining ring and produce sidewall leakage. If back-pressure is used, the ring should be rigid or the outside of the ring should be pressurized to keep it from expanding. It is impossible to measure the B coefficient of the soil in a fixed-wall permeameter and confirm saturation by ensuring that the B coefficient is close to unity. However, the degree of saturation of the test specimen can be determined after the test has been terminated. Sidewall leakage can be identified by using a double-ring permeameter (Fig. 1). With the double-ring device, the outflow from the central portion of the test specimen is collected separately from the outflow near the sidewall. It is convenient to size the ring that separates the inner and outer collection areas such that the inner and outer rings have the same area. This is accomplished if the diameter of the separating ring is 0.707 times the diameter of the test specimen. Thus, if the volume of effluent liquid collected over a period of time is the same in the inner and outer rings, there is no sidewall leakage. On the other hand, if more effluent liquid is collected from the outer ring than from the inner ring, it is likely that sidewall leakage is affecting the results of the test. If there is a large difference between flow rates in the inner and outer rings, the test results are invalid.

The advantages of a double-ring permeameter are illustrated by the following example. A sample of kaolinite was mixed to a moisture content of 34 percent, which is 2 percent wet of optimum from a standard Proctor compaction test. The soil was then compacted into a 10-cm-diameter compaction mold following ASTM Standard D698, Method A (commonly known as "standard Proctor"). The soil was permeated with water (0.01 N $CaSO_4$) under a hydraulic gradient of 100. The hydraulic conductivity quickly stabilized at approximately 7 x 10^{-8} cm/s, although the outflow from the outer ring was 10 percent higher than from the inner ring. This may have been the result of sidewall leakage, or the soil near the sidewalls of the permeameter may have been slightly more permeable for reasons that have nothing to do with sidewall leakage. After permeation with water, the permeant liquid was switched to heptane, a neutral non-polar organic solvent that is immiscible with water. Heptane has been reported to cause large increases in the hydraulic conductivity of compacted clay [1, 10]. The rate of inflow quickly increased when heptane was introduced, and the test was continued for about 4 pore volumes of flow. For the last 2 pore volumes of flow, 1,200 ml of outflow went into the outer ring while 10 ml went into the inner ring. Virtually all of the flow appeared to be going down the sidewall of the permeameter. The sidewall flow was undoubtedly caused by shrinkage of the soil when the heptane was introduced.

Critical Importance of Stress

The hydraulic conductivity of clay is controlled by the distribution of pore sizes and the characteristics of the diffuse double layers that surround clay particles. The pore size distribution may be altered by changing the stress applied to the soil. In compacted clay, there is evidence that a few relatively large pores conduct most of the liquid that passes through the compacted soil. If the largest voids can be significantly compressed, the hydraulic conductivity can be altered significantly. Some representative data are shown in Fig. 2. In this case,

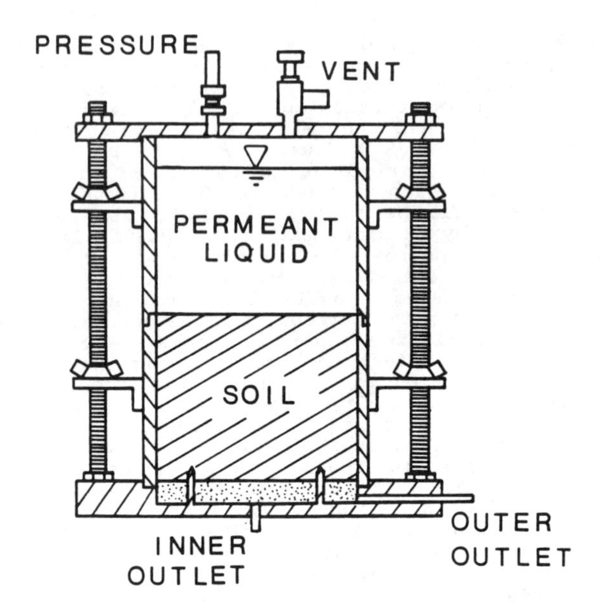

Figure 1. Double-ring permeameter.

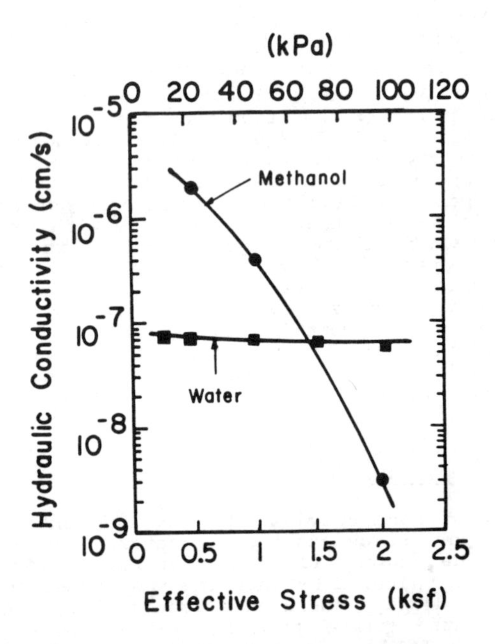

Figure 2. Influence of effective stress upoon the hydraulic conductivity of compacted kaolinite permeated in a consolidation-cell permeameter with either water or methanol (from Foreman and Daniel, 1986).

separate specimens of compacted kaolinite were permeated at low hydraulic gradient with either methanol or water. The tests were performed initially at low effective stress, and then the effective stress was increased in increments. With methanol, 2 pore volumes of permeant liquid were forced through the soil before the stress was increased. The permeameter was a consolidation-cell device in which the soil was permeated under conditions of constant head. Methanol tends to cause soil to shrink, and the added stress causes consolidation and a reduction in hydraulic conductivity.

Boynton and Daniel [5] tested specimens of compacted clay that were either cracked or uncracked. The cracked specimens were trimmed from blocks of compacted clay that had been allowed to desiccate at room temperature in the laboratory for 1 to 3 days. The specimens were permeated under backpressure in a flexible-wall permeameter at low hydraulic gradient and variable effective stress. Results for one series of tests are shown in Fig. 3. The cracked specimen underwent a large reduction in hydraulic conductivity as the effective stress was increased above roughly 4 psi (30 kPa).

It is very difficult to perform hydraulic conductivity tests in the laboratory and maintain low effective stress. The effective stress varies throughout the test specimen but is greatest at the effluent end of the test specimen (Fig. 4). For the case of zero external confinement, the seepage pressures alone are sufficient to produce the following effective stresses in a 3-inch (75 mm) thick specimen:

Hydraulic Gradient	Effective Stress at Effluent End psf	kPa
1	25	1
10	165	8
100	1,570	75
1,000	17,140	821

If any external effective stress is applied (as is always the case in flexible-wall permeameters, for instance), the values in the table shown above would be even higher.

Because increases in effective stress cause reductions in hydraulic conductivity, it is best not to use effective stresses in laboratory tests that are larger than field values. As the table shown above indicates, one cannot perform hydraulic conductivity tests at high hydraulic gradient and simultaneously maintain a low effective stress.

IN-SITU HYDRAULIC CONDUCTIVITY TESTS

Several methods are available for measuring the hydraulic conductivity of compacted clay liners in the field. Five methods of testing will be discussed in this section:

1. Borehole tests.
2. Porous probes.
3. Air-entry permeameters.
4. Lysimeter pans.
5. Ring infiltrometers.

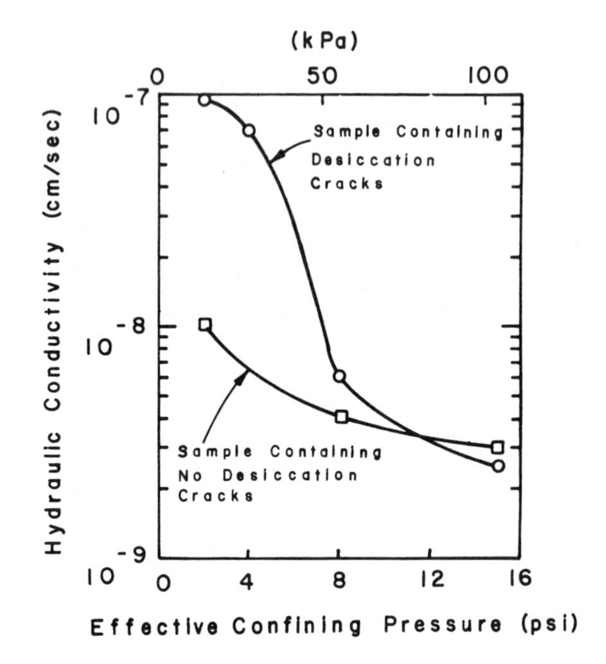

Figure 3. Influence of effective stress upon hydraulic conductivity of compacted fire clay (from Boynton and Daniel, 1985).

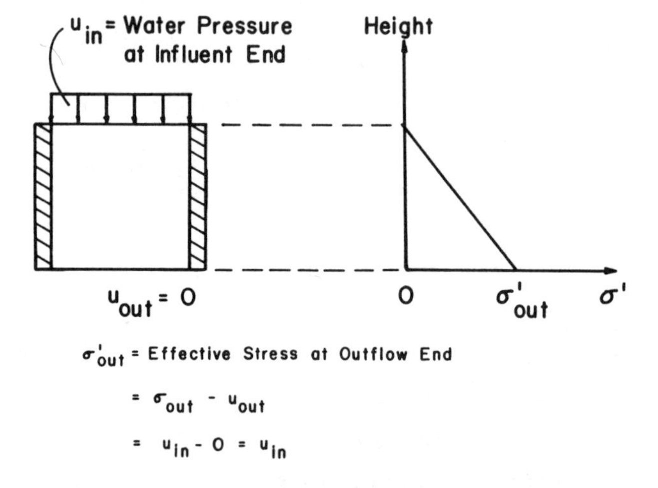

Figure 4. Variation in effective stress with position in a soil sample that is permeated in a rigid-wall permeameter with no externally-applied stress. The weight of the sample has been assumed to be negligible.

Borehole Test

A two-stage, borehole hydraulic conductivity test was developed by Boutwell and is described by Boutwell and Derick [4]. The apparatus is shown schematically in Fig. 5. The device is installed by drilling a hole to depth Z. The depth Z must be at least 5 times larger than D to avoid ambiguities in the interpretation of test results. Further, the depth from the base of the borehole to the bottom of the liner should at no time be less than about 5D for the same reason. After the hole is drilled, a casing in placed inside the hole and the annular space between the casing and borehole is sealed with grout. A cap is placed on the permeameter. A reservoir is used to fill the casing and standpipe.

Once the permeameter has been assembled and filled with water, the Stage I tests are initiated. The elevation of zero pore water pressure in the soil is assumed to be at the base of the casing such that the head driving flow is H as shown in Fig. 5a. A series of falling-head tests are performed, and the hydraulic conductivity from Stage I (k_1) is computed as follows:

$$k_1 = \frac{\pi d^2}{11\, D(t_2 - t_1)} \ln(H_1/H_2) \tag{1}$$

The values of k_1 are plotted as a function of time (e.g., Fig. 6) until steady conditions are reached, which typically seems to take from a few days to as much as 2 to 3 weeks. The steady-state value of k_1 is used for subsequent calculations.

Next, the top of the permeameter is removed and the hole is deepened with an auger or by pushing a thin-walled sampling tube. The ratio of length to diameter (L/D) in the uncased zone (Fig. 5b) should be between 1.0 and 1.5. The permeameter is reassembled, and a series of falling head tests are again performed. The head loss (H) is assumed to be as shown in Fig. 5b. The hydraulic conductivity from Stage II (k_2) is calculated as follows:

$$k_2 = [A / B] \ln(H_1/H_2) \tag{2}$$

where:

$$A = d^2 \{\ln[(L/D) + (1+(L/D)^2)^{1/2}]\}$$

$$B = 8\, D\, (L/D)\, \{1 - 0.562 \exp[-1.57(L/D)]\}$$

The values of k_2 are plotted as a function of time until k_2 ceases to change significantly (Fig. 6).

Next one picks arbitrary values of m, where m is defined as:

$$m = [k_h / k_v]^{1/2} \tag{3}$$

(k_h and k_v are the hydraulic conductivities in the horizontal and vertical directions, respectively) and calculates the corresponding values of k_2/k_1 from the expression:

$$k_2/k_1 = \frac{\ln\left\{(L/D) + \left[1 + (L/D)^2\right]^{1/2}\right\}}{\ln\left\{(mL/D) + \left[1 + (mL/D)^2\right]^{1/2}\right\}} \tag{4}$$

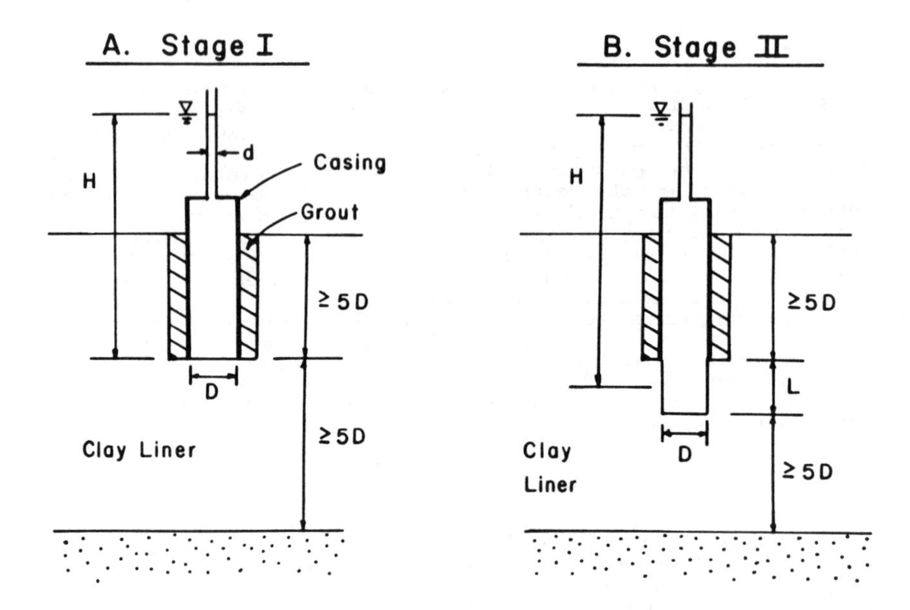

Figure 5. Two-stage, borehole permeability test (Boutwell and Derick, 1986).

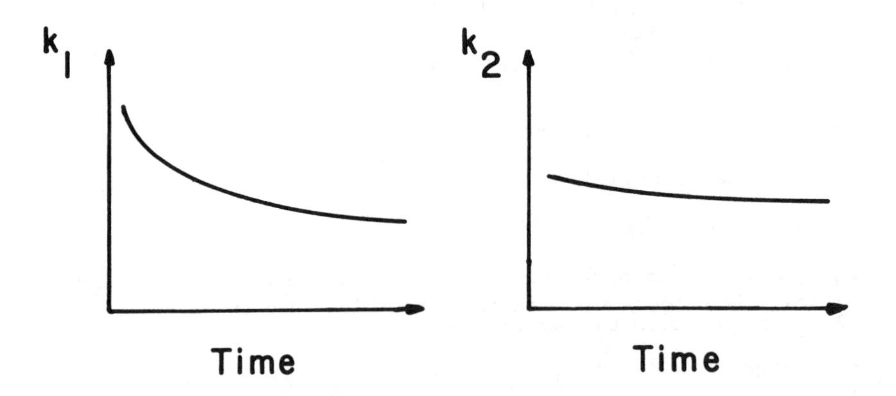

Figure 6. Variation of k_1 and k_2 with time.

Typically, values of m ranging from 1 to as much as 10 might be used. The resulting data (values of k_2/k_1 and corresponding values of m) are plotted, e.g., as shown in Fig. 7 for L/D = 1.0 and L/D = 1.5.

The next step is to determine the actual value of k_2/k_1 from the series of falling-head tests that were performed and the values of k_1 and k_2 determined from Eqs. 1 and 2. From the graph of k_2/k_1 vs. m (Fig. 7) that had been prepared, one determines the actual value of m that corresponds to the actual value of k_2/k_1. The hydraulic conductivities in the vertical and horizontal directions are computed as follows:

$$k_h = m\, k_1 \tag{5}$$

$$k_v = (1/m)\, k_1 \tag{6}$$

The critical assumptions that were made are that the soil is homogeneous (the same soil is permeated in Stages I and II), pore water pressure in the surrounding soil is zero at the base of the permeameter (Stage I) or the center of the uncased section (Stage II), the soil that is permeated is sufficiently far removed from any boundaries of the liner that the test results are unaffected by boundary conditions, the degree of saturation of soil through which water flows is uniform (which is analogous to assuming that the soil is essentially fully saturated in the vicinity of the test), effects of soil suction are negligible, steady conditions have been reached in the two stages, the soil undergoes no volume change during a falling-head test, and the equations developed by Hvorslev [13] for flow around cased boreholes are correct.

The advantages of borehole tests are that the devices are relatively easy to install, they can be installed at great depth, the cost is relatively low, the hydraulic conductivity in both the vertical and horizontal directions can be measured, and relatively low hydraulic conductivities (as low as about 1×10^{-9} cm/s) can be measured. The disadvantages are that the effects of incomplete and variable saturation are unknown, the influence of soil suction upon the results is ill defined, the test cannot be used near the top or bottom of a liner, and the volume of soil that is permeated is relatively small. Boutwell and Derick [4] indicate that the test has worked well and present several case histories.

Porous Probes

Porous probes may be pushed or driven into a clay liner (Fig. 8). Water is then introduced, and the rate of flow is measured. The test may be a constant-head test or a falling-head test. Hvorslev's [13] equations are usually employed to determine hydraulic conductivity. For a saturated, homogeneous, isotropic soil, the equation for a constant-head test is:

$$k = (q\, C)/(2\, \pi\, L\, H) \tag{7}$$

where:

$$C = \ln \{ (L/D) + [1 + (L/D)^2]^{1/2} \}$$

For a falling-head test:

$$k = \frac{d^2\, C}{8\, L\, (t_2 - t_1)} \ln (H_1/H_2) \tag{8}$$

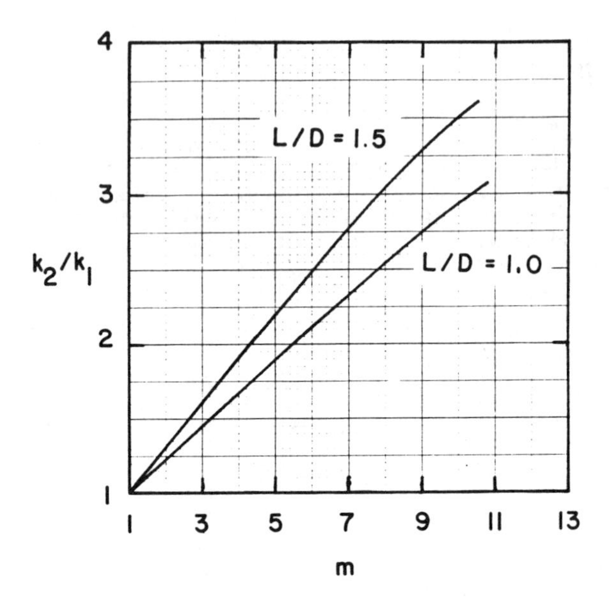

Figure 7. k_2/k_1 versus m for L/D = 1.0 and L/D = 1.5.

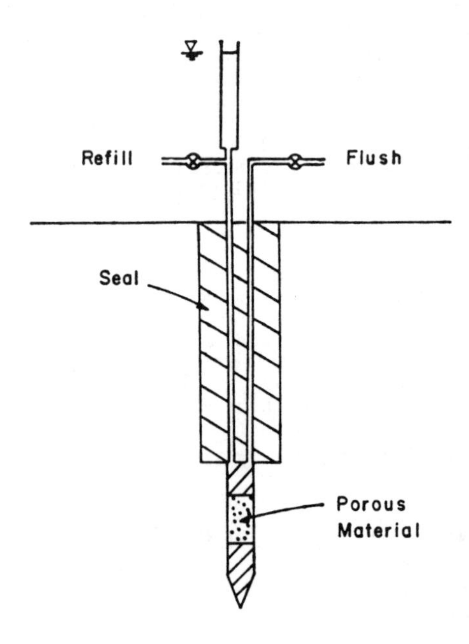

Figure 8. Porous probe.

Torstensson [19] has developed a cone-tipped, porous-probe permeameter that is especially convenient for hydraulic conductivity testing at depth. A porous probe is pushed into the soil beneath the bottom of a bore hole, and then casing is brought to the surface. A chamber is lowered down the casing and brought into contact with the porous probe using a hypodermic needle and septa. The chamber contains both air and water. The air in the chamber is pressurized (or evacuated) to any desired pressure. As water flows out of (or into) the probe, the air pressure in the chamber changes. A pressure transducer monitors the pressure changes. The quantity of flow and heads are computed from Boyle's Law and the measured change in the gas pressure in the chamber. When the test is complete, the chamber is raised out of the casing with a septum providing a down-hole seal on the probe.

The important assumptions with porous probes are that the pore water pressure in the soil that surrounds the probe is known, the soil that is permeated is far removed from any boundaries that might influence the results, the degree of saturation does not vary spatially in the zone in which flow occurs, the soil is incompressible, and Hvorslev's equations are correct. Advantages and disadvantages are similar to those of borehole tests, except that the probes are easier to install and that flow is almost entirely horizontal. On the negative side, the porous probes tend to be small and therefore to permeate a relatively small volume of soil. Also, the soil may be smeared when the probe is installed.

Air-Entry Permeameter

The air-entry permeameter (AEP) was developed by Bouwer [2, 3]. The permeameter, shown in Fig. 9, consists of a ring with a cover. The ring is typically about 1 to 2 ft (30 to 60 cm) in diameter. The ring (without the top) is either pushed into the soil, driven into the soil, or sealed into the soil with a grouted trench (Fig. 9). The top is then bolted to the ring and the device is filled with water.

The AEP is used in two stages. In the first stage, the device is operated as a sealed, single-ring infiltrometer. A buret, or standpipe, on top of the device is used to measure the rate of infiltration (I):

$$I = q/A \tag{9}$$

where q is the volume of flow per unit time and A is the area of the ring ($\pi D^2/4$, where D is the inside diameter of the ring). The infiltration rate is measured and plotted versus time. Typically with clay liners, I drops rapidly in the first few hours of a test and then slowly declines over a period of many days or weeks. However, the infiltration rate that is observed a few hours into a test, after the large drop in I is over, is typically used. Thus, Stage I would normally last from a few hours to perhaps as much as several days. Bouwer recommends a falling-head test to determine I, but the author believes that falling-head tests are ill advised in this application because gas bubbles and the AEP itself will tend to contract as the head falls. A constant head test, e.g., with a Marriotte device, is recommended.

When the first stage is complete, a valve to the buret (standpipe) is closed. At this point, the AEP is sealed. A negative pressure develops in the AEP as the unsaturated soil tries to suck water out of the AEP. The negative pressure that develops in the AEP is measured with a gauge on the top plate. When the minimum water pressure reading is obtained, the test is complete.

When soils are dried from an initially saturated condition, a water content-suction curve of the type shown in Fig. 10 may be observed. When an unsaturated soil is moistened, the soil is typically still unsaturated even when the suction is zero because air bubbles are entrapped. The water-entry suction (p_w) and air-entry suction (p_a) are denoted in Fig. 10. According to Bouwer, the minimum water pressure during the second stage of testing is reached when the air-entry value of the wetted zone is reached. At that point, air will start moving upward through the wetted zone. As soon as the minimum water pressure is read on the gauge, the AEP is disassembled and the depth to the wetting front (L_f) is measured. Typically, this depth is no more than a few centimeters.

The air-entry suction at the edge of the wetting front (p_a') is defined as follows:

$$p_a' = -u_w - (L_f + G) \cdot \gamma_w \tag{10}$$

where u_w is the minimum water pressure (a negative value) measured with a pressure gauge located a distance G above the ground surface (Fig. 10) and γ_w is the unit weight of water. Bouwer suggests, based on experience with various soils (but not compacted clay), that the water-entry suction at the wetting front (p_w') is approximately one-half the air-entry suction. Bouwer assumes that the suction at the wetting front is p_w'. The pressure head at the wetting front (h_{pf}) is assumed to be:

$$\begin{aligned} h_{pf} &= -p_w'/\gamma_w \\ &= -\,1/2\ p_a'/\gamma_w \\ &= -\,1/2\ [-u_w - (L_f + G)\,.\,\gamma_w\,]/\gamma_w \\ &= 1/2\ \{(u_w/\gamma_w) + (L_f + G)\} \end{aligned} \tag{11}$$

From Darcy's Law, hydraulic conductivity is computed as follows:

$$\begin{aligned} k &= q/(i\ A) \\ &= (q/A)/(h/L_f) \\ &= (I\ L_f)/(H + L_f - h_{pf}) \end{aligned} \tag{12}$$

where I is the rate of infiltration from the first stage of testing (Eq. 9), L_f is the depth of wetting front determined after the test is complete, h is the head loss across the wetted zone, H is the pressure head at the ground surface inside the AEP during the first stage of testing (infiltration), and h_{pf} is the pressure head at the base of the wetting front (Eq. 11).

Several important assumptions are made: (1) the suction at the base of the wetting front is the water-entry value (p_w') of the soil; (2) the water-entry value is one-half the air-entry value; (3) the water pressure read on the gauge during the second stage of testing is (when corrected for elevation) the negative of the air-

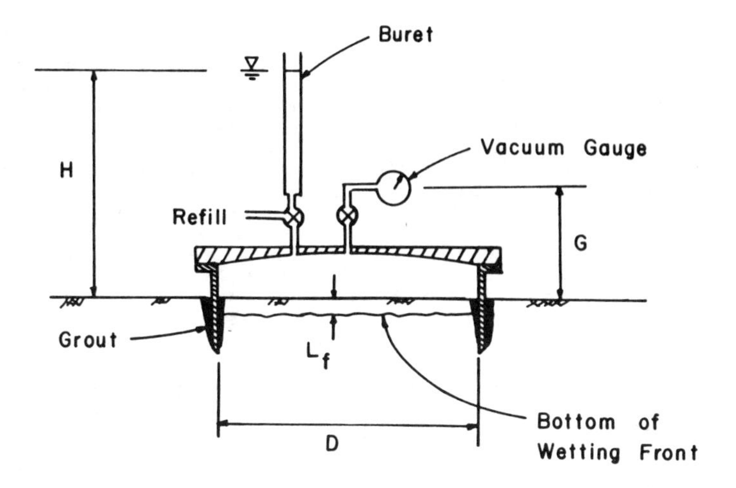

Figure 9. Air-entry permeameter (after Bouwer, 1978).

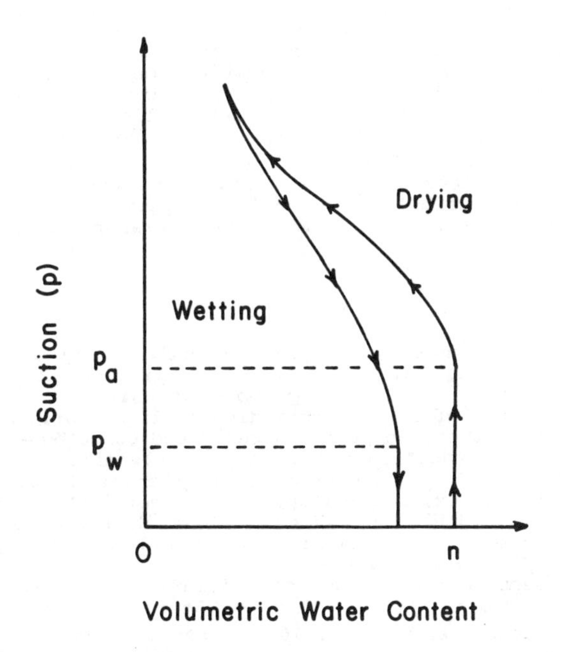

Figure 10. Definition of water-entry suction (p_w) and air-entry suction (p_a) (after Bouwer, 1978).

entry suction (p_a'); (4) the AEP is completely rigid; (5) the soil is incompressible; and (6) the air pressure in the soil beneath the wetting front is atmospheric. These assumptions are unproven and in some cases unlikely to be correct. Also, at the wetting front, there is a very large change in suction over a short distance. Water pressures are positive just above the wetting front and negative just below it. An argument could be made for assuming that the pressure head at the base of the wetting front is zero.

The advantages of the air-entry permeameter (AEP) are relatively rapid measurements, the permeation of a relatively large area, and assurance that k in the vertical direction is measured. The disadvantages are that several simplifying assumptions have been made but are unverified for compacted clay, the depth of soil tested is shallow, and very low infiltration rates are difficult to measure because of thermal effects and the compliance of AEP itself. Knight and Haile [15] used the AEP on an earthen liner and reported k's of 5×10^{-9} to 3×10^{-7} cm/s. Tests in the laboratory on "undisturbed" samples produced k's that averaged about one-half an order of magnitude less than k's from the AEP.

Lysimeter Pans

Lysimeter pans (Fig. 11) are placed beneath a clay liner to collect seepage over a limited area. The pans can be constructed of any reasonably impervious material, but geomembrane materials are especially convenient. The pan is backfilled with sand, gravel, or synthetic drainage materials. Hydraulic conductivity is calculated from the measured rate of flow into the pan and Darcy's law.

One problem with pan lysimeters is that substantial time may be needed before reasonably steady conditions develop (many weeks or months for materials having low k). Also, disadvantages include a need to install the lysimeter pan before the liner is constructed, a need for gravity drainage of the discharge pipe, and difficulty in measuring very low k's (less than about 1×10^{-8} cm/s). The advantages of lysimeter pans include few experimental or theoretical ambiguities, the possibility of testing very large volumes of soil, and the ability to use tracers to study transport of chemicals through a liner.

Infiltrometers

Infiltrometers may be single- or double-ring devices. In addition, the rings may be open infiltrometers or sealed infiltrometers (Fig. 12). The purpose of using two rings is to restrict the amount of lateral spreading of liquid originating from the inner ring. If the seepage can be forced to be essentially one dimensional, then reduction of data is greatly simplified. If the width (or diameter) of the ring is many times larger than the thickness of the liner, lateral spreading is of little consequence. However, if the width of the ring is less than 3 to 5 times the thickness of the liner, significant spreading can occur and a double-ring device is preferable.

With respect to open versus sealed infiltrometers, open rings are much easier to construct. However, it is difficult to measure infiltration rates less than 1×10^{-7} cm/sec (2.6 mm per month) with open rings because such small changes in water level are difficult to measure and because evaporative losses can be much greater than infiltration. For clay liners, sealed infiltrometers are preferable.

Daniel and Trautwein [8] describe a sealed double-ring infiltrometer (SDRI). Several improvements have recently been made. The SDRI (Fig. 13) consists of a sealed fiberglass inner ring with a

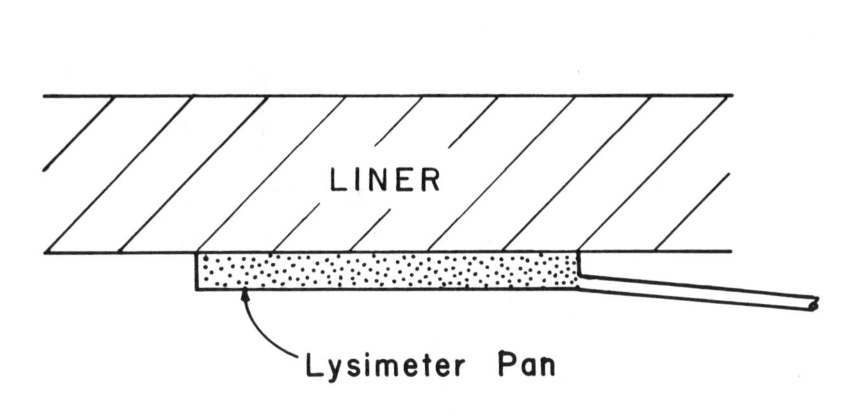

Figure 11. Lysimeter pan.

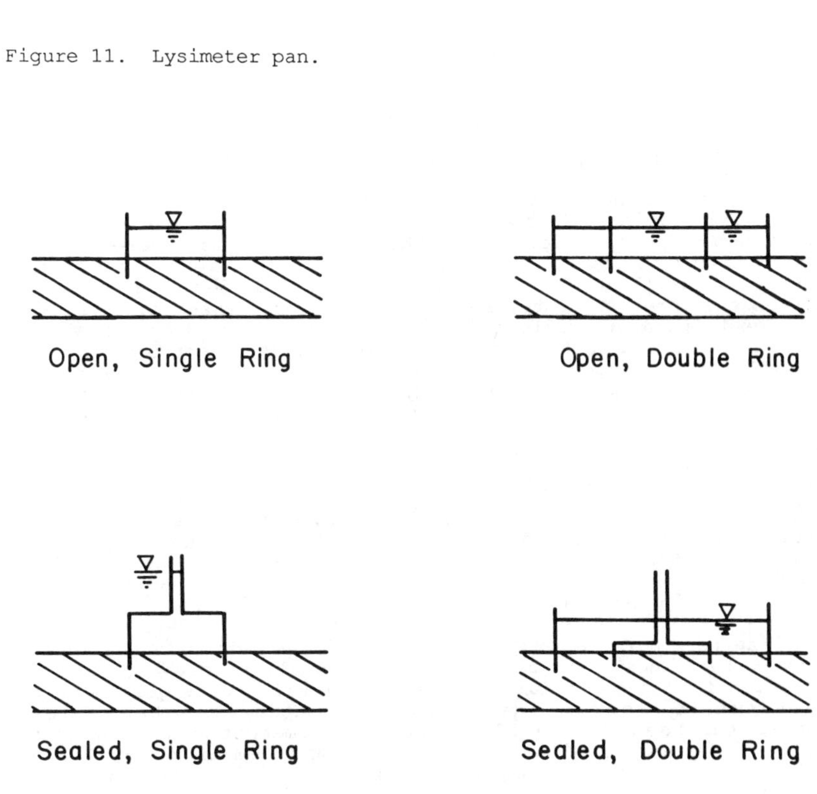

Figure 12. Infiltrometers.

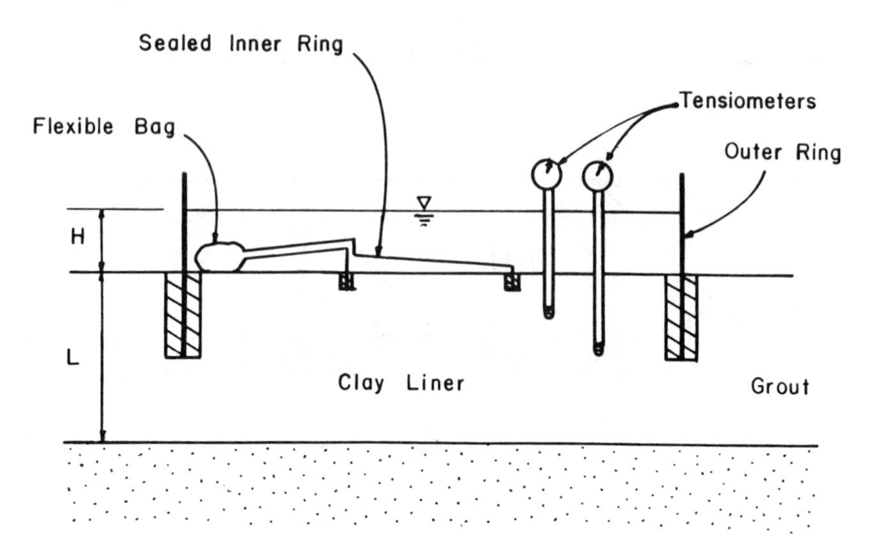

Figure 13. Sealed double-ring infiltrometer.

width of 5 ft (1.5 m) and an open metallic outer ring that has a width of 12 ft (3.7 m). Both rings are embedded into trenches that are sealed with a suitable grout. It is best to install tensiometers at several depths to monitor the progress of the wetting front. The rings are filled such that air is removed from the inner ring. A small flexible bag is filled with water, weighed, and attached to the inner ring. The entire SDRI is covered with a tarpaulin and periodically the bag is removed, weighed, and (when necessary) refilled. The bag ensures that the differential pressure between the inner and outer ring is always zero (assuming there are no large temperature differences between inner and outer rings), so that expansion and contraction of the inner ring cannot occur.

The infiltration rate (I) from the sealed inner ring is calculated from Eq. 9, where the rate of flow (q) over an interval of time (t) is determined by occasionally weighing the bag attached to the inner ring. The hydraulic conductivity (k) is:

$$k = I/i \tag{13}$$

where i is the hydraulic gradient. It is best to wait until the wetting front reaches the bottom of the liner, in which case the gradient (i) is:

$$i = (H + L)/L \tag{14}$$

where the terms are defined in Fig. 13. This equation is only valid if the water pressure is zero at the base of the liner. If there is a suction from the underlying material, i will be larger than the value given by Eq. 14. Regardless of whether the wetting front reaches the base of the liner, the tensiometers can be used to measure the gradient above the wetting front if two porous probes at different depths are attached to a differential pressure gauge or manometer.

A photograph of the sealed inner ring is shown in Fig. 14. The outer, metallic ring is assembled at the site (Fig. 15). The entire SDRI, with tensiometers installed, is shown in Fig. 16. The flexible bag is shown in Fig. 17.

Figure 14. Sealed fiberglass inner ring. (Courtesy of Trautwein Soil Testing Equipment Co.)

Figure 15. Outer ring prior to assembly. (Courtesy of Trautwein Soil Testing Equipment Co.)

Figure 16. Inner ring, outer ring, and tensiometers installed. The buckets on sealed inner ring serve as weights to avoid accidental uplift. (Courtesy of Trautwein Soil Testing Equipment Co.)

Figure 17. Flexible bag used to measure rate of infiltration from inner ring. (Courtesy of Trautwein Soil Testing Equipment Co.)

The assumptions made in using the SDRI are that the seepage beneath the inner ring is one dimensional, the liner is underlain by freely-draining material with known (preferably negligible) suction, flow is steady, any swelling of the soil is complete, and the average temperature difference between water in the inner and outer rings is the same. The advantages of the SDRI are that a large volume of soil is tested, there are few experimental ambiguities if the test is continued to steady state, seepage is vertical, and there is essentially no disturbance of the soil. Disadvantages are that the installation time is greater than with most other methods, the test may need to last a relatively long time (30 to 90 days is typical with clay liners, and even longer times may be needed in some cases, e.g., liners with low initial degrees of saturation), there cannot be a material of lower hydraulic conductivity at the base of the liner and extremely low k's (less than 1×10^{-8} cm/s) cannot be measured accurately.

The advantages of testing a large volume of soil are so compelling that, in the author's opinion, the pan lysimeter and SDRI are the preferred in-situ methods for measuring the hydraulic conductivity of clay liners. Pan lysimeters and SDRI's enjoy most of the same advantages; the SDRI just measures inflow into the liner and the lysimeter measures outflow from the liner.

CONCLUSION

In this paper, methods for measuring the hydraulic conductivity in both the laboratory and the field have been discussed. It was seen that no testing method is without simplifying assumptions and limitations. For laboratory tests, the key consideration is to avoid excessive confining stress. In-situ tests range from small, easy-to-use devices to much larger, less convenient apparatuses. No one type of in-situ test will always be the best. However, one of the objectives of an in-situ hydraulic conductivity test should be to permeate a representative volume of soil. The pan lysimeter and sealed double-ring infiltrometer (SDRI) are best able to test a large volume of soil.

REFERENCES

1. Anderson, D. C. "Does Landfill Leachate Make Clay Liners More Permeable?" Civil Engineering, 52(9):66-69 (1982).

2. Bouwer, H. "Rapid Field Measurement of Air-Entry Value and Hydraulic Conductivity of Soil as Significant Parameters in Flow System Analysis," Water Resources Research 2:729-732 (1966).

3. Bouwer, H. Groundwater Hydrology (New York: McGraw-hill, 1978), p. 480.

4. Boutwell, G. P., and R. K. Derick "Groundwater Protection for Sanitary Landfills in the Saturated Zone," paper presented at Waste Tech '86, National Solid Waste Management Association, Chicago, IL (1986).

5. Boynton, S. S., and D. E. Daniel "Hydraulic Conductivity Tests on Compacted Clay" Journal of Geotechnical Engineering 111(4):465-478 (1985).

6. Daniel, D. E. "Predicting Hydraulic Conductivity of Clay Liners," Journal of Geotechnical Engineering 110(2):285-300 (1984).

7. Daniel, D. E., Trautwein, S. J., and D. McMurtry "A Case History of Leakage from a Surface Impoundment," in Proceedings of Seepage and Leakage from Dams and Impoundments, ASCE, (1985), pp. 220-235.

8. Daniel, D. E., and S. J. Trautwein "Field Permeability Test for Earthen Liners," in Proceedings of In-Situ '86, ASCE Specialty Conference in Blacksburg, VA (New York: S. P. Clemence, ed,, 1986), pp. 146-160.

9. Day, S. R., and D. E. Daniel "Hydraulic Conductivity of Two Prototype Clay Liners," Journal of Geotechnical Engineering 111(8):957-970 (1985).

10. Foreman, D. E., and D. E. Daniel "Permeation of Compacted Clay with Organic Solvents," Journal of Geotechnical Engineering 112(7):669-681 (1986).

11. Goodall, D. C., and R. M. Quigley "Pollutant Migration from Two Sanitary Landfill Sites near Sarnia, Ontario," Canadian Geotechnical Journal 14(2):223-236 (1977).

12. Griffin, R. A., et al. "Mechanisms of Contaminant Migration through a Clay Barrier," in Proceedings of the Eleventh Annual Research Symposium, Land Disposal of Hazardous Waste, U.S. EPA, Cincinnati, Ohio (1985), pp. 27-38.

13. Hvorslev, M. J. "Time Lag in the Observation of Ground-Water Levels and Pressures," U. S. Army Engineers Waterways Experiment Station, Vicksburg, MS(1949).

14. Keller, C. K., van der Kamp, G., and J. A. Cherry "Fracture Permeability and Groundwater Flow in Clayey Till near Saskatoon, Saskatchewan," Canadian Geotechnical Journal 23:229-240 (1986).

15. Knight, R. B., and J. P. Haile "Construction of the Key Lake Tailings Facility," in Proceedings of the International Conference on Case Histories in Geotechnical Engineering, St. Louis, MO (1984).

16. Reades, D. W., Pohland, R. J., Kelly, G., and S. King, Discussion, Journal of Geotechnical Engineering, in press (1986).

17. Rogowski, A. S. "Hydraulic Conductivity of Compacted Clay SOils," in Proceedings of the Twelfth Annual Research Symposium, Land Disposal, Remedial Action, Incineration, and Treatment of Hazardous Waste, U.S. EPA, Cincinnati, Ohio, (1986) pp. 29-39.

18. Rowe, R., and J. R. Booker "1-D Pollutant Migration in Soils of Finite Depth," Journal of Geotechnical Engineering 111(4):479-499 (1985).

19. Torstensson, B. A. "A New System for Ground Water Monitoring Review," Ground Water Monitoring Review 4(4):131-138 (1984

GEOMEMBRANE/SYNTHESIZED LEACHATE COMPATIBILITY TESTING

Jon W. Hughes, Senior Engineer,
AWARE Incorporated, West Milford, NJ

Michael J. Monteleone, Geotechnical Engineer,
AWARE Incorporated, West Milford, NJ

INTRODUCTION

This liner-waste compatibility study was conducted to determine the most suitable geomembrane material for the lining of a proposed hazardous waste landfill. The information contained within this report has been developed to meet the liner-waste compatibility requirements of 40 Code of Federal Regulations (CFR) 270.21 and 40 CFR 264.301(A).

Two sets of exposed flexible membrane liners; 50 mil chlorinated polyethylene (CPE), and 80 mil high density polyethylene (HDPE), were evaluated after 0 (unexposed), 30, 60, 90 and 120 days.[1] Liners were exposed to water, non-solvent and solvent. The one-sided immersion tub incubation procedure was used for the exposure.[2] The liners were evaluated by a series of the following tests.

- o mechanical (tensile strength, strain at failure, modulus at 10% strain, toughness, tear resistance and puncture strength)
- o diffusion (water vapor transmission and radioactive tracer diffusion)

The results of the mechanical tests indicate that the HDPE is generally more stable over the 120-day test period than the CPE. This, however, was not a clear-cut decision because, in a few instances (particularly in the solvent), the HDPE also showed data scatter. This data scatter is not unusual in tests of this type, and could have been caused by an error in the tests or the test methods themselves.

Geotechnical and Geohydrological Aspects of Waste Management, D. J. A. van Zyl et al., Eds., © 1987 Lewis Publishers, Inc., Chelsea, Michigan – Printed in USA.

In the diffusion tests, the HDPE clearly out performed the CPE. Both water vapor transmission and diffusion through the exposed liners were more uniform and much lower in the HDPE than in the CPE.

Overall (certainly weighed by the visually deteriorated CPE samples in solvent for 90 and 120 days), the test results indicate that the HDPE liner is the recommended geomembrane for this particular application.

SELECTION OF SYNTHESIZED LEACHATE

The composition of the synthesized leachate used in the liner testing program was based on facility manifest logs from 1982 through 1985. Two waste solutions were used to evaluate liner performances under different conditions. The first solution (Waste 1) was a non-solvent solution, made of the constituents shown in Table 1. The second solution (Waste 2) was a mix of the non-solvent used in Waste 1 and the solvent constituents shown in Table 2.

Table 1. Waste 1 - Non-solvent Solution

EPA Waste ID No.	Category	Approximate % of Total	Amount Added 500 ml (ml)	Amount Added 55-gal (gal)
D001	Ignitible	10.5	53	6.0
D002	Corrosive Lagoon	1.0	5	0.5
D008	Lead	3.0	5	1.5
D009	Mercury	0.2	1	0.1
D013	Lindane	73.0	365	40.0
F006	Wastewater Treatment Sludges	3.4	17	2.0
F007	Plating Bath	1.0	5	0.5
	Oil Lagoon	3.0	1	1.5
	Scrubber Dust	4.8	2	2.5

Table 2. Waste 2 - Solvent Solution

EPA Waste ID No.	Category	Approximate % of Total	Total Vol. Added to 500 mil container Laboratory	55-gal drum Field
D001	Ignitable	8.8	53	5.0
D002	Corrosive Lagoon	0.8	5	0.5
D008	Lead	2.5	15	1.5
D009	Mercury	0.2	1	0.1
D013	Lindane	60.8	365	34.0
F001	Halogenated Solvents	5.0	30	2.75
F002	Halogenated Solvents	5.0	30	2.75
F003	Non-Halogenated Solvents	5.0	30	2.75
F006	Wastewater Treatment Sludges	2.8	17	1.5
F007	Plating Bath	0.8	5	0.5
	Oil Lagoon	2.5	15	1.5
	Scrubber Dust	4.0	24	2.5
U226	1,1,1-Trichloroethane	0.4	2.5	0.25
U239	Xylene Mixture	0.4	2.5	0.25
U154	Methanol	0.4	2.5	0.25
U019	Benzene	0.4	2.5	0.25

Most of the chemical components used in these two solutions were obtained by sampling wastes from the containers at the site. Laboratory grade samples were used for three wastes; Mercury, 1,1,1-Trichloroethane and Methanol, not available at the site. The sampling was performed by using; glass sampling rods, scoops, and drum hand pumps.

To prevent any unexpected reactions during the mixing of the synthesized waste in the field, a 500 mil sample of Waste 1 was mixed in a laboratory under a well-ventilated hood. Each chemical was added, in order of final proportions, and any reactions or effects were noted. Three actions were noted during the mixing of the non-solvent wastes: an increase in solution temperature (when Oil Lagoon waste was added), fluctuation of the pH (around 5.5), and the formation of a small gas cloud.

The solvents were then added proportionally to the synthesized waste. No extensive reactions were noted. It was, however, difficult to maintain a completely mixed waste due to the immiscibility of the oil, water and organic materials. These materials separated into different phases.

The "full scale" consolidation of the designed leachate was conducted in a two-step process at the facility. In the first step, Waste 1 was composited into two 55-gallon drums. The wastes were added in proportional order to the drums. Agitation was accomplished by a drum mixing rod.

Next, the solvents in Waste 2 were mixed in two plastic 5-gallon containers. Samples of the non-solvent and solvent wastes were tested by the facility's contract laboratory.

METHOD OF EXPOSURE

The liners were placed in 11½" by 13½" by 5½" Rubber Maid tubs, arranged in two 8' x 4' x 6" constructed wooden boxes. Waste 1 was added to all of the lined tubs designated for leachate testing. Water was added to the lined tubs designated for control. Solvents were added to the designated tubs and thoroughly mixed. The wooden lids of the liner testing boxes were closed and locked.

The lined tubs were checked weekly for evaporation of the leachate and water. When evaporation of waste occurred, the tubs were refilled from the original mix of wastes. This resulted in the liners being exposed to higher concentrations then shown in Tables 1 & 2. Evaporation from the water-filled tubs was replaced with tap water. Samples of liners were taken after 30, 60, 90 and 120 days, and the tests described below were conducted.

During this operation, a fungi growth was found to be on the surface of the following exposures: 30, 60 and 90 days HDPE Waste 2 exposures and the 30, 60, 90 and 120 days CPE Waste 1 exposures. This fungi was sampled and identified as Fusarium sp., by the contract laboratory.

The 30, 60, 90 and 120-day samples were removed from the tubs, packaged and shipped by overnight service directly to the testing laboratory. There they were removed from the shipping cartons and individually stored in plastic bags until testing. Samples were kept in a constant temperature (70°F ± 5°) and humidity (50% ± 10%) room. The samples were removed from the bags and cut into test specimens as per EPA Method 9090 or the appropriate ASTM Standard.

MECHANICAL TESTS AND RESULTS

A series of mechanical tests were performed on unexposed and exposed liners in both the machine direction (MD) and cross, or transverse, machine direction (XMD).[3] They were tested in replicates using a computerized, constant-rate-of-extension Instron 4062 testing machine with automatic recording equipment. The strain rate was constant at the prescribed value during each test. The grips for the tension test were ASTM D1682 - Grab Tensile Test clamps which allow for gripping over the entire face of the sample specimen. Test specimen shape for tension properties was a "dogbone" configuration conforming to EPA's Method 9090.

Tensile Strength

Tensile strength was determined by taking the failure load of the test specimen divided by its width (0.25") and its respective nominal thickness (0.050" for CPE and 0.080" for HDPE). Five tests each in the MD and XMD for each exposure time were performed, averaged and plotted. The response for the 0, 30, 60, 90 and 120 day results are given in Figures 1, 2 and 3 for water, non-solvent and solvent respectively. Although the CPE liner at 90 and 120 days was noticeably deteriorated by the solvent, the least deteriorated areas were tested. The results for both water and non-solvent are approximately constant with the CPE slightly lower in strength than the HDPE. More important, the solvent results in both materials show large variations, the CPE demonstrating very low strength at high exposure time.

Strain at Failure

Strain at failure was calculated by dividing the deformed specimen length at failure by its original length, yielding a result in percentage. Decreases in this value indicate an embrittlement sometime caused by leaching of the liner's plasticizer. The responses are shown on Figures 1, 2 and 3, for water, non-solvent and solvent, respectively. The largest variations are in the CPE in solvent, which could indicate a leaching phenomenon.

Modulus

The slope of the initial portion of the load versus deformation curve is the modulus of elasticity, a representation of the stiffness of the material. Since the curve is never completely linear, this value was calculated at 10% strain. The units are shown in pounds per inch. Division by the liner's thickness will give the conventional stress units. Figures 1, 2 and 3 give the results for water, non-solvent and solvent, respectively. In this set of response curves, the CPE was more consistent than the HDPE, particularly in the solvent waste exposure.

Toughness

Toughness is a measure of the combined effects of strength and ductility. It is indicative of the energy a material can absorb before it fails. Toughness is one half the failure stress, multiplied by failure strain. Results are shown on Figures 4, 5 and 6, for water, non-solvent and solvent, respectively. Units for these figures are measured in stress values since the strain is used as a ratio. These figures show that the water and non-solvent had little effect, but that the solvent influenced the CPE in a rather dramatic manner. Toughness is seen to decrease uniformly in both the MD and XMD for the CPE in Figure 6.

Tear Resistance

Test specimens of the liners were taken to determine the tear resistance as per EPA's Method 9090. This property relates to the resistance to propagation of a tear initiated at a 90° cut in the center of the liner. The liner is gripped at the ends and pulled in a tension testing machine at the prescribed rate. Results, provided in pound resistance at the maximum value, are shown in Figures 4, 5 and 6, for the water, non-solvent and solvent respectively. Results for water and non-solvent are both quite stable, but for the CPE in solvent at 90 and 120 days the results were essentially zero. In these cases the liner was visably deteriorated and effected by the solvent.

Puncture Strength

In this test, the liner is supported horizontally in a test fixture and a plunger is penetrated through it at a rate of 10 inch/min. This test, an adapted form of ASTM D3787, with the ball changed to a 5/16" blunt ended metal rod, is under draft standard status by Committee D35 on Geotextiles and Geomembranes. The maximum force required for complete penetration is reported in pounds. Results are given in Figures 4, 5 and 6, for water, non-solvent and solvent, respectively. Note that there is no MD and XMD given, since orientation is not a factor. Results show that water and non-solvent incubation were non-influential, but the solvent did influence the CPE. At 90 and 120 days the puncture strength fell to essentially zero.

DIFFUSION TESTS AND RESULTS

"Micro-structure" assessment is an alternate concept of looking at the influence of liquid incubation on liner compatibility. By diffusing a gas or liquid through the liner specimen, the rate of molecular movement can be measured and compared to the as-received material. Note that this is a fundamental departure from mechanical testing whereby the "macro-structure" is addressed, using an entire test specimen to obtain its average strength. In diffusion tests, a tiny channel through the liner should markedly influence the results.

Water Vapor Transmission

In this ASTM E96 test, the liner test specimen is fixed and sealed over a water-filled aluminum cap and placed in a controlled relative humidity chamber. The difference in relative humidity of 100% within the cup and the chamber (usually held at about 40%) causes water vapor to pass through the liner test specimen. Weight loss of the assembly with time is monitored to calculate the water vapor transmission (WVT), permeance, or vapor permeability. WVT was used in this study to determine relative differences only. However, the higher the WVT, the greater is the diffusion through the liner test specimen. Figure 7, gives

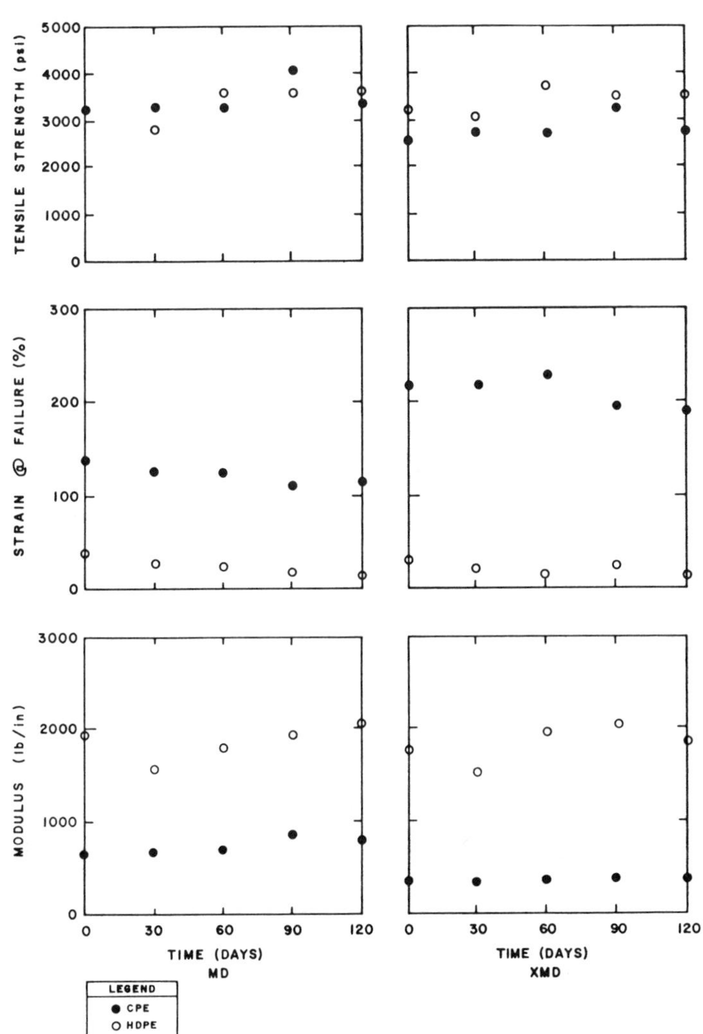
FIGURE I
TENSILE STRENGTH, STRAIN AT FAILURE AND MODULUS OF CPE AND HDPE INCUBATED IN WATER
TENSILE STRENGTH (psi)
STRAIN @ FAILURE (%)
MODULUS (lb/in)
TIME (DAYS)
MD
TIME (DAYS)
XMD
LEGEND
CPE
HDPE

FIGURE 2

TENSILE STRENGTH, STRAIN AT FAILURE AND MODULUS OF CPE AND HDPE INCUBATED IN NONSOLVENT

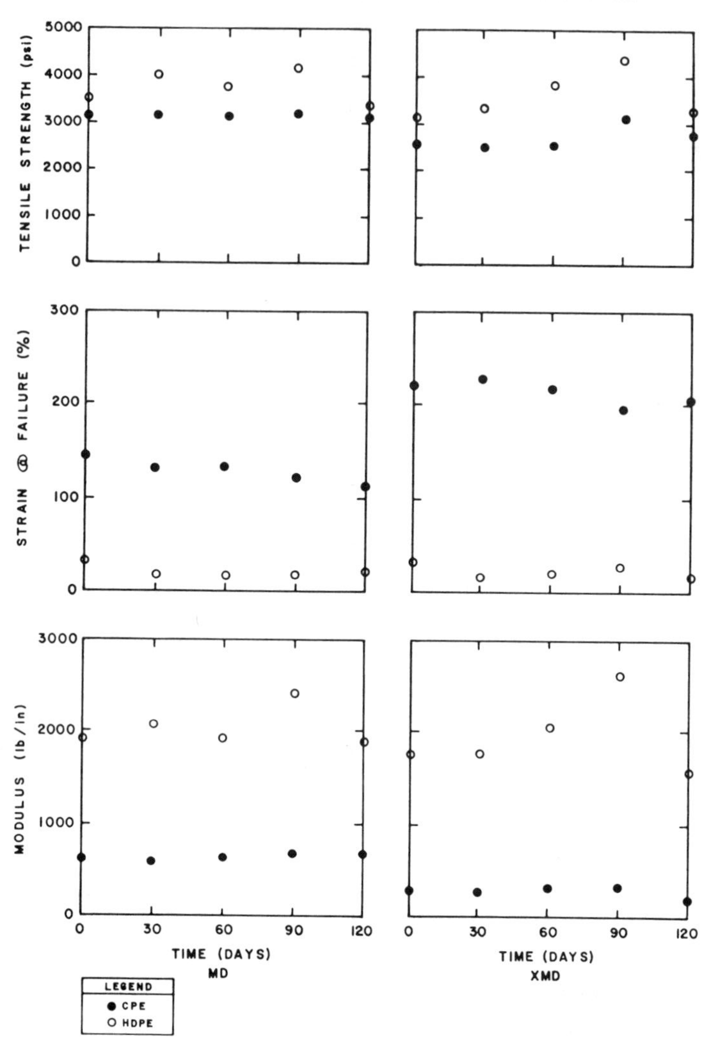

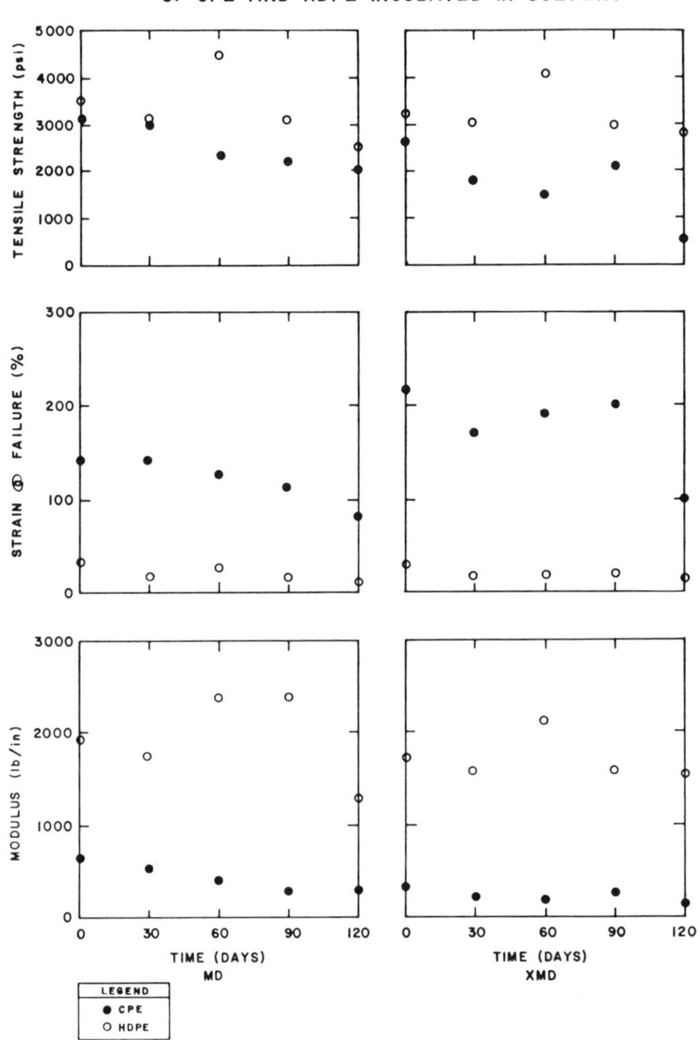
FIGURE 3
TENSILE STRENGTH, STRAIN AT FAILURE AND MODULUS OF CPE AND HDPE INCUBATED IN SOLVENT
TENSILE STRENGTH (psi)
5000
4000
3000
2000
1000
0
STRAIN @ FAILURE (%)
300
200
100
0
MODULUS (lb/in)
3000
2000
1000
0
0
30
60
90
120
TIME (DAYS)
MD
XMD
LEGEND
CPE
HDPE

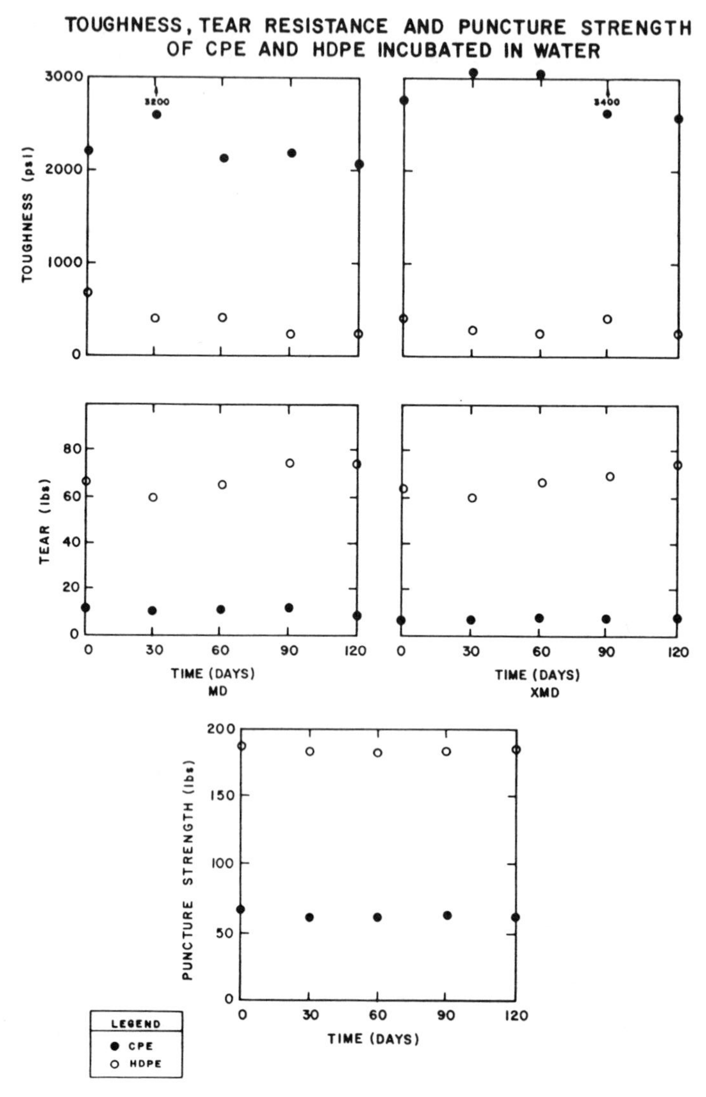
FIGURE 4
TOUGHNESS, TEAR RESISTANCE AND PUNCTURE STRENGTH OF CPE AND HDPE INCUBATED IN WATER
TOUGHNESS (psi)
3000
2000
1000
0
3200
3400
TEAR (lbs)
80
60
40
20
0
0
30
60
90
120
TIME (DAYS)
MD
TIME (DAYS)
XMD
PUNCTURE STRENGTH (lbs)
200
150
100
50
0
TIME (DAYS)
LEGEND
CPE
HDPE

FIGURE 5

TOUGHNESS, TEAR RESISTANCE AND PUNCTURE STRENGTH OF CPE AND HDPE INCUBATED IN NONSOLVENT

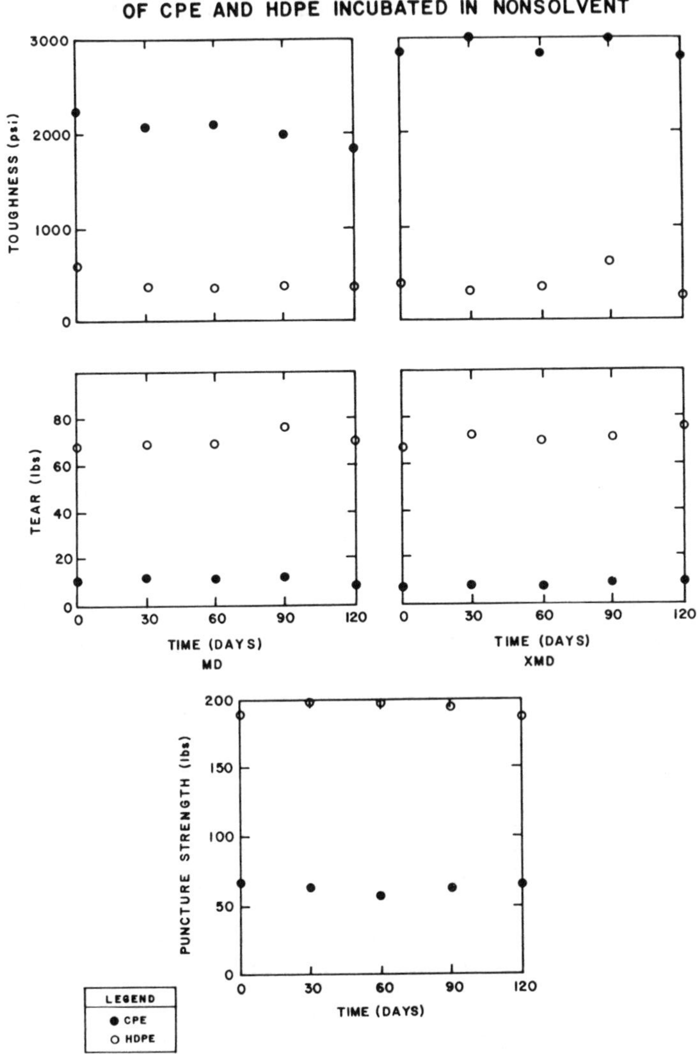

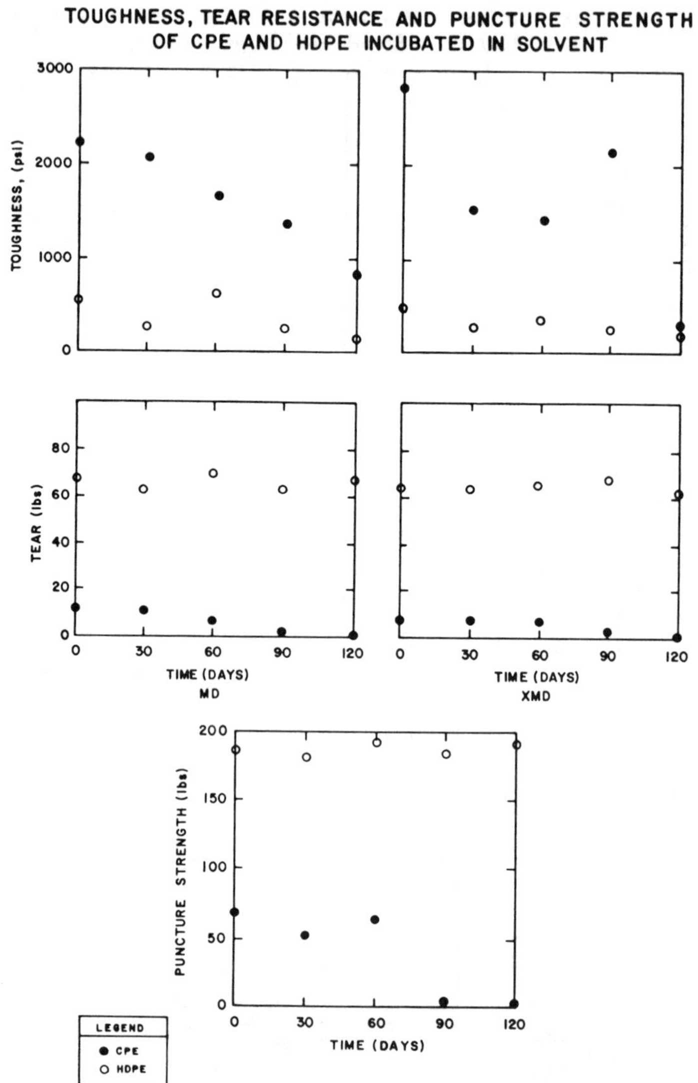
FIGURE 6
TOUGHNESS, TEAR RESISTANCE AND PUNCTURE STRENGTH OF CPE AND HDPE INCUBATED IN SOLVENT
TOUGHNESS, (psi)
3000
2000
1000
0
TEAR (lbs)
80
60
40
20
0
0
30
60
90
120
TIME (DAYS)
MD
XMD
PUNCTURE STRENGTH (lbs)
200
150
100
50
LEGEND
CPE
HDPE

FIGURE 7

WATER VAPOR TRANSMISSION RESULTS OF CPE AND HDPE IN WATER, NONSOLVENT AND SOLVENT

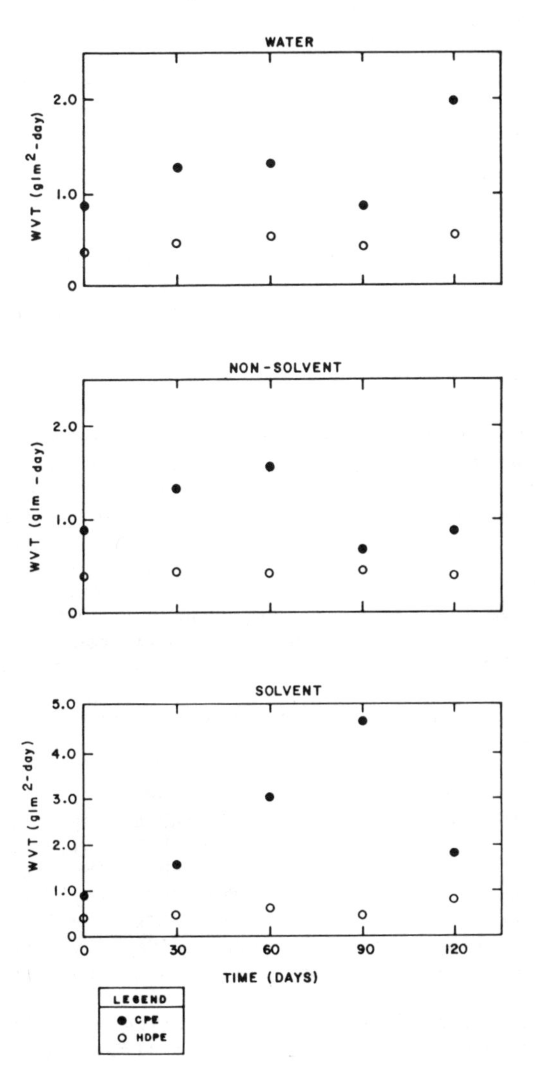

the WVT responses for water, non-solvent and solvent, incubated liners. It is clear that CPE has consistently higher WVT values than does HDPE, varying considerably with exposure. Note that there was sufficient non-deteriorated solvent treated CPE material to do 90 and 120 day tests. The response was very erratic as seen on the graph.

Radioactive Tracer Diffusion

This alternate diffusion type test measures a radioactively-tagged liquid on top of the liner test specimen as it diffuses through the liner over time. By using a Geiger counter on the underside of the liner one can "count" the amount of liquid diffusing through it and calculate a diffusion coefficient. Using benzene with Carbon 14 as the tagged radioactive isotope, the data of Figure 8, was generated. Here, on a greatly reduced scale, the CPE is seen to be 10 times higher in diffusion than the HDPE. Furthermore, there is a gradual increase in the solvent treated CPE with increasing exposure time.

SUMMARY AND CONCLUSIONS

The study of CPE versus HDPE flexible membrane liners measured a series of water, non-solvent and solvent, incubated test samples. Samples tested after incubation times at 30, 60, 90 and 120 days were added to the as-received material to give five points on the test response curve. The tests performed on the liner specimens fell into two broad classifications: mechanical (tensile strength, strain at failure, modulus at 10% strain, toughness, tear resistance and puncture strength) and diffusion (water vapor transmission and radioactive tracer diffusion). Impact tests were not possible due to insufficient test sample material. The results of this large testing program were averaged and plotted on Figures 1 through 8. In general it was found that:

tensile strength	CPE ≅ HDPE	
strain at failure	CPE > HDPE	(by 5 to 10 times)
10% modulus	CPE < HDPE	(by 2 to 3 times)
toughness	CPE > HDPE	(by 5 to 10 times)
tear resistance	CPE < HDPE	(by 3 to 7 times)
puncture strength	CPE < HDPE	(by 3 to 7 times)
water vapor transmission	CPE > HDPE	(by 2 to 6 times)
radioactive tracer diffusion	CPE > HDPE	(by 8 to 10 times)

While this information is interesting for liner design purposes, the trends and/or stability of these test values over the 120-day test period is of maximum concern. Here it was observed that the HDPE usually out performed the CPE. This is particularly significant for the solvent incubated CPE, which was very erratic in its test behavior and visably deteriorated in the 90 and 120-day samples.

FIGURE 8

RADIOACTIVE TRACER DIFFUSION COEFFICIENT RESULTS OF CPE AND HDPE IN WATER, NONSOLVENT AND SOLVENT

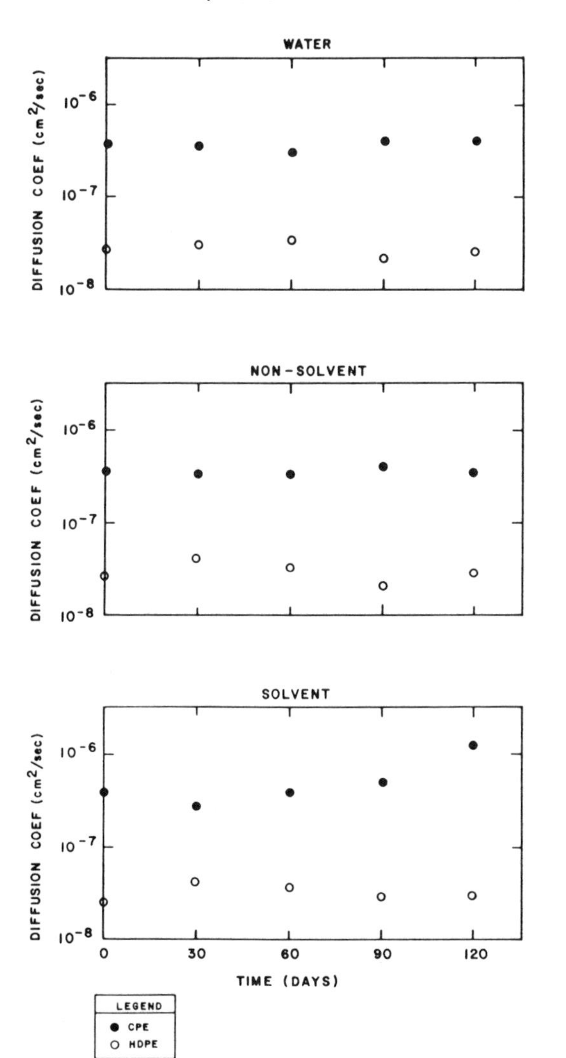

Perhaps the most telling set of test data are the diffusion results shown on Figures 7 and 8. While these diffusion tests are not usually mandated via EPA Method 9090, they were performed in this testing study. Both WVT, and radioactive tracer diffusion values, clearly show that water and benzene movement through the CPE, is significantly greater than through HDPE, under all conditions evaluated.

References

1. U.S.EPA, "EPA Method 9090 - Compatibility Tests for Waste and Membrane Liners," Office of Solid Wastes, Washington, DC, 1984.

2. Lord, A. E. and Koerner, R. M., "Fundamental Aspects of Chemical Degradation of Membranes," Proc. Intl. Conf. on Geomembranes, Denver, Colorado, June 1984, pp. 293-298, Published by IFAI, St. Paul, Minnesota.

3. Koerner, R. M., Designing with Geosynthetics, Prentice-Hall Publ.Co., Englewood Cliffs, NJ, 1986.

4. Monteleone, M. J., Puncture and Impact Behavior of Various Geosynthetics, Drexel University, Masters Thesis, June 1986.

INVESTIGATION OF CONTAMINATION AT A LANDFILL IN WISCONSIN

Jey K. Jeyapalan, Joe Winch, and Wesley S. Ethiyajeevakaruna
University of Wisconsin, Madison, Wisconsin

INTRODUCTION

The landfill site addressed in this paper is located in Wisconsin. A gravel pit at this site was utilized by a tire manufacturing corporation of Wisconsin during the period of 1962 to 1966 for dumping wastes generated. The surrounding private wells have been found to be contaminated with several chemicals and a hydrogeological investigation was conducted in 1982 by a local engineering company for the manufacturing corporation. Although this previous study provided some information about the level of contamination and a groundwater monitoring program around the waste pits, the consequences of the waste contamination at this site and at the nearby areas were not addressed in detail. A detailed investigation was undertaken by the authors for studying the contamination at this landfill and the results of this investigation are given in this paper. Various remedial action and site cleanup procedures were also recommended to indicate possible alternatives for aquifer restoration at this site, but these aspects of the study will be reported in another paper elsewhere.

DESCRIPTION OF THE CONTAMINATION PROBLEM

Geometry of the Site

The landfill site is located in Southeast Wisconsin. The approximate plan areas of the three waste pits are 5 acres, and 2 at 2 acres. The approximate depth of these pits based on previous investigation is less than 15 feet and the average depth to the bedrock is about 50 feet. The distance of hijgh capacity wells from the waste pits is about 1200 feet and there are several other private wells even closer than 1200 feet to the three waste pits.

Geotechnical and Geohydrological Aspects of Waste Management, D. J. A. van Zyl et al., Eds., © 1987 Lewis Publishers, Inc., Chelsea, Michigan—Printed in USA.

Materials Dumped

The solid waste survey by the manufacturer presented a summary of materials dumped at this site. The materials related to the manufacturing processes employed at this plant at that time were also known. Other records of the manufacturer indicated that the predominant materials were Xylol Sludge, MEK sludge, Xylol decants, and small amounts of other wastes. One of the drums found in the pit was analyzed as part of the investigation and was found to contain primarily Xylene, Tetrahydrofuran, Bromoform, and other organics. Based on the estimate by the U.S. EPA and the assumption that most durms disposed in these waste pits had Xylol type wastes, estimates of various organic chemicals dumped initially at this site were computed during this investigation and these are summarized in Table 1.

Table 1. Types of Chemicals found at the Landfill

Trans-1,2-Dichloroethene	0.7
Vinyl Chloride	0.1
Tetrahydrofuron	3.6
M,O,P Xylene	73.7
Dichlorodifluoromethane	0.1
Acetone	0.4
Tetrachloroethylene	0.7
N-Butyl Acetate	0.7
Tricholorethylene	0.7
Toluene	0.4
Methyl Ethyl Ketone	0.3
Methyl Isobutyl Ketone	0.7
Bromoform	8.1
Ethylbenzene	0.6

Recorded Consequences

The known consequences of dumping of these wastes at this site include contamination of groundwater at several private wells and undesirable health effects to occupants in the vicinity of these waste pits. This investigation however, will focus only on the groundwater contamination aspects of this site.

SUBSURFACE CONDITIONS

Site Geology

The bedrock geology below the landfill is predominantly of the cambrian formation of sandstone with some dolomite and rock. The thickness of unconsolidated material around the site is typically 0 to 100 feet. The types of ice age deposits forming Dane County soils, particularly in the study area are outwash and till of moraines, providing a high coarse aggregate potential. The grain size curves determined for these materials during the 1956-1961 county groundwater survey indicated high water yielding capacity of various aquifers in these soils. The specific capacities of wells in the area were producing up to as high as 30,000 gal/day for a depth of 50 feet. Furthermore, this survey also defined the regional hydrogeology indicating the major groundwater divides and the site is located in the Yahara River system.

Soil Surveys

Soil survey maps for the Dane County have been in use since 1918 and the latest map for which the field work was completed before 1960 provides ample valuable information about the subsurface conditions and land use classifications. The shapes of the gravel pits after gravel and sand production had depleted the source were also clearly shown on this map. When the outer peripheries of the three waste pits and the remaining area behind the pits are combined todgether, this total area matches the shape and size of the gravel pit. This shows that the soil survey was completed when the gravel pit was unoccupied by the wastes dumped by the manufacturer and therefore, sufficient information was available at that time from this soil survey about the soil conditions at this dite. The soil survey map indicates that there are primarily seven types of soils involved at the site and that there is severe to very severe danger of contaminating groundwater if sanitary landfills are located in this area. The soil survey also recommends that for all sanitary landfills deeper than 5 or 6 feet, on site studies of the underlying strata, water table, and hazard of aquifer pollution and drainage into groundwater need to be made. The landfills involved in this study were deeper than 6 feet and the summary of soil survey data for the site indicates that the seasonable high watertable is only 0 to 5 feet deep from the natural ground surface.

Permeability Data

The permeability data obtained from baildown slug tests at several locations ranges from 1.7×10^{-5} to 3.0×10^{-3} cm/sec. These values however, do not represent the results one would obtain for permeability of more permeable soils present at this site. Based on the classifications of soils shown on the Geological Survey for the site, the permeability values can be estimated as 0.1 to 0.05 cm per sec.

DESCRIPTION OF CONTAMINATION LEVELS

Analyses of groundwater samples from many of the water quality wells and private wells indicated contamination to various degrees with a number of organics. All available data from various water quality testing programs were organized during this investigation to observe any possible trends. The details are presented in the sections below.

Variation in Contaminant Levels

Typical variations in Xylene and Tetrahydrofuran for one of the many wells are shown in Figure 1. With the exception of variation of Xylene at one well, in all the other cases, the concentration of the contaminants Xylene, Tetrahydrofuran, and Ethylbenzene increased with time during the period of July 82 to January 85. The amounts of Toluene, Chlorobenzene, and Carbon Tetrachloride had also been on an increasing trend during the period of July 82 to March 85.

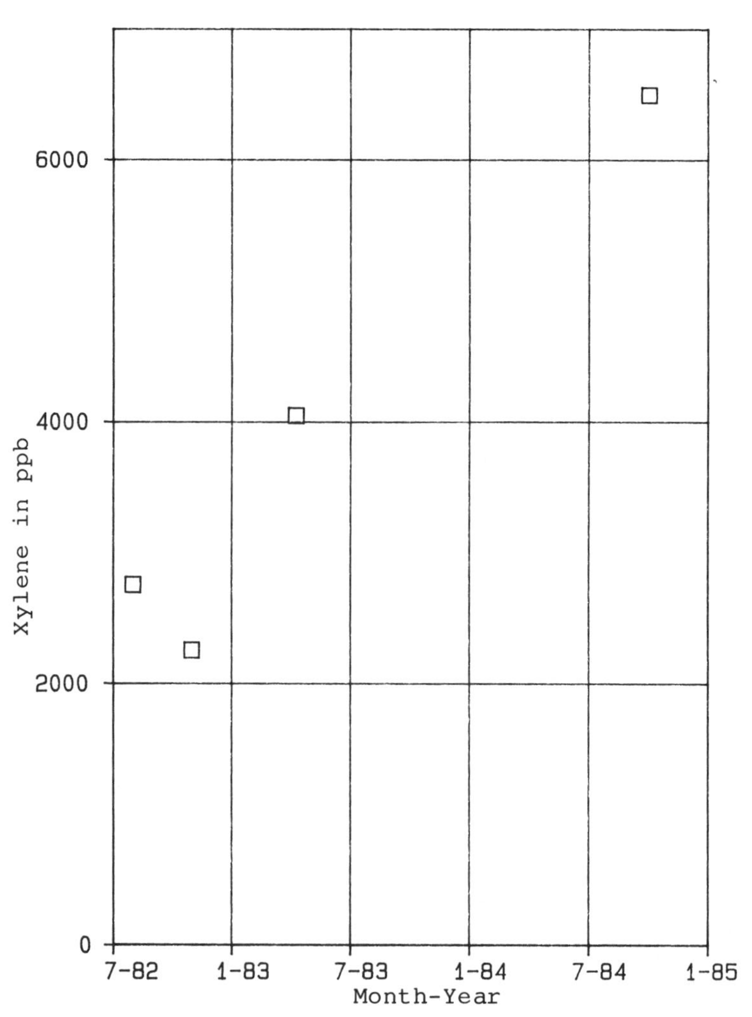

Figure 1A. Variation of Xylene at One Well

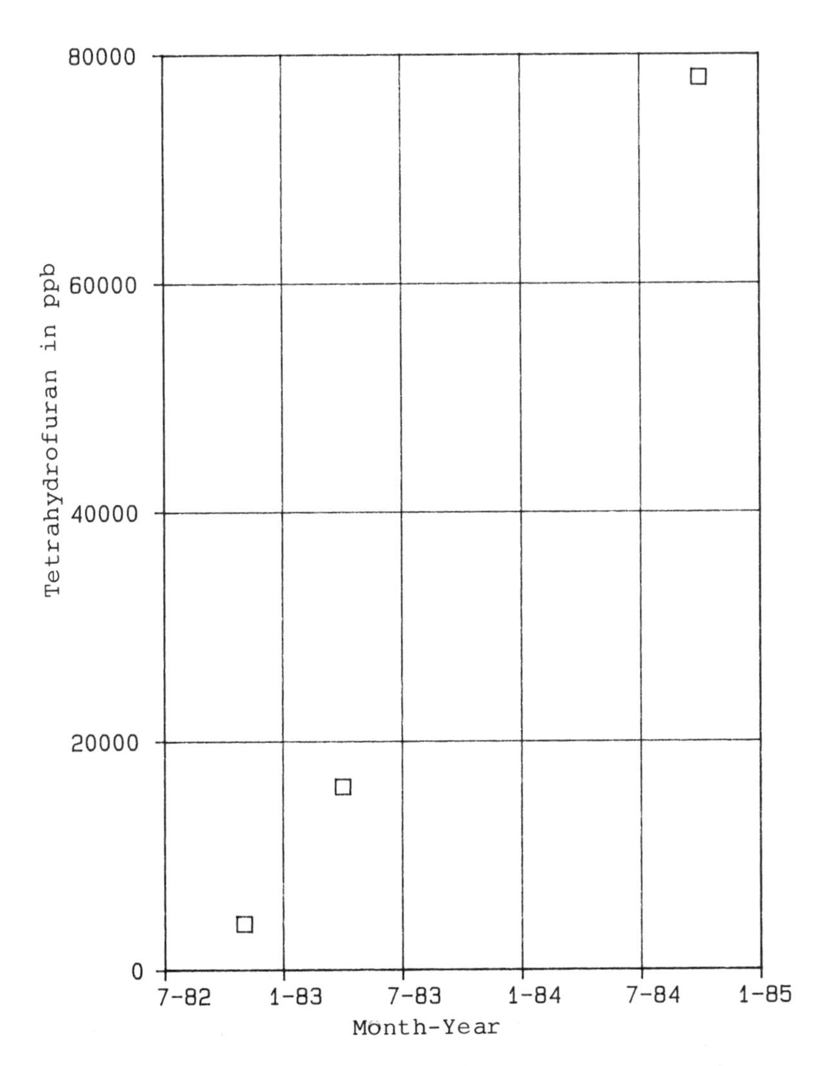

Figure 1B. Variation of Tetrahydrofuran at One Well

Variation in pH Values

The groundwater resources survey conducted during 1956-1960 by U.S.G.S. obtained some data on the quality of groundwater at various locations in the Dane County. The wells tested on 1-9-58 and 4-19-60 provide the very data needed to establish the quality of groundwater around the site before wastes were dumped. The ranges of pH as given by Dane County soil surveys, and U.S.G.S. before waste dumping at this site for groundwater are summarized in Table 2 with those from laboratory tests performed during the last five years on groundwater quality. In all cases, the pH of the groundwater around this site has become more acidic due to the presence of organics with pH lower than 7.0.

Table 2. pH of Groundwater at the Landfill Site before and after Dumping of hazardous Wastes

Well No.	Before Dumping (Soil Survey)	Before Dumping (Geol. Survey)	After Dumping
1	7.4-7.8	7.4-7.9	6.9
2	7.9-8.4	7.4-7.9	6.9
3	7.9-8.4	7.4-7.9	7.5
4	7.9-8.4	7.4-7.9	7.6
5	7.9-8.4	7.4-7.9	6.9
6	7.4-7.8	7.4-7.9	7.3
7	7.4-7.8	7.4-7.9	6.9-7.5
8	7.4-7.8	7.4-7.9	6.8-7.4
9	7.9-8.4	7.4-7.9	-
10	7.4-7.8	7.4-7.9	6.8-7.4
11	7.4-7.8	7.4-7.9	6.9-7.6
12	6.1-7.8	7.4-7.9	7.0-7.4

Variations in Conductivity values

The variations in conductivity values for the groundwater at a typical private well are shown in Figure 2. The groundwater quality survey of 1956-60 by U.S.G.S. gave a representative conductivity value of 642 micro-mhos/cm for the groundwater around the site in 1958. Thus, the drastic variations in conductivity at most of the above private wells can be attributed to the contamination.

Degree of Compliance with Wisconsin Regulations

The variations in contaminant levels, pH, and conductivity indicate that the groundwater supply is contaminated by the chemicals dumped at the site and that the contaminant plume is still active. The sensitivity of the state-of-the-art analytical methods vary from one instrument to another and from one technique to another. The U.S. EPA in its technical resource document-SW846 has reported the detection limits for various organics. The limits used by Wisconsin DNR are in all cases higher than those used by the U.S. EPA. According to NR 140 of DNR requirements, the water quality test laboratory shall utilize the analytical methodology specified in rules or approved by the U.S. EPA. Where no analytical methodology is specified, the laboratory shall use an analytical methodology with a limit of detection and limit of quantification below the preventive action limit. Where the limit of detection or limit of quantification is above the preventive action limit for that substance, the laboratory shall use the best available analytical methodology to produce the lowest limit of detection and limit of quantification.

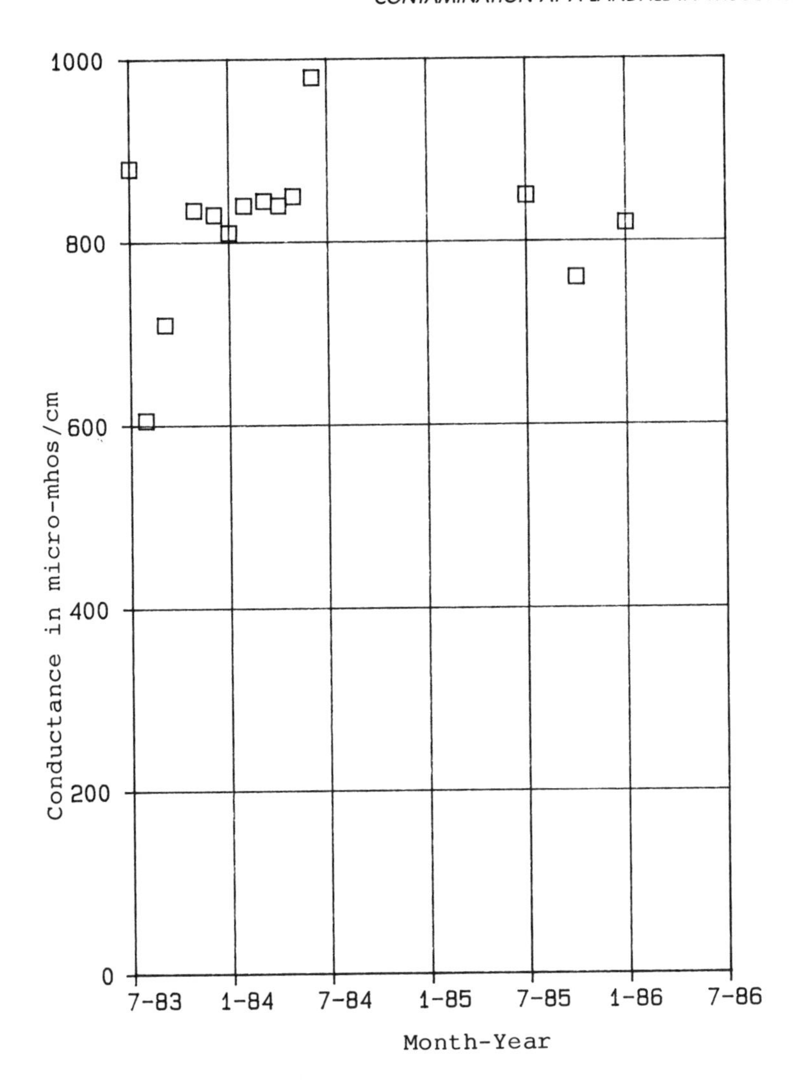

Figure 2. Variation of Conductivity at one of the Wells at Site

A careful review of the analytical laboratory data indicated that the detection limits of the procedures used by the test laboratories for determining the presence of various organics in the groundwater from various wells are significantly higher than those required by DNR and by U.S. EPA. Therefore, the claims made by these labs that the contamination levels were below detection limits did not preclude the potential of not complying with those requirements of NR 140. Therefore, based on the data available from the tests one cannot claim that when the substance was BDL that there was no contamination.

DESCRIPTION OF GROUNDWATER CONDITIONS

The flow of groundwater is a fundamental requirement for contaminants to travel considerable distances and therefore a correct interpretation of the hydrogeology of the site is necessary in order to be able to predict the extent of contamination. This section will focus on the characteristics of groundwater flow at this site.

Representative Precipitation and Infiltration

The climatological data for the period of 1951 to 1980 were used to determine the average precipitation over the 30 year period for each month of the year. The rate of infiltration controls the amount of recharge available for groundwater supply at any site and this is controlled mostly by the type of soil near the surface of the ground. The infiltration rate varies with the time lapsed after the beginning of the precipitation. Using this approach, the representative infiltration rates for all months of the year during the 30 year period were calculated.

Effects of Mounding

During and after a recharge activity, the groundwater level will rise below the waste pit due to the inflow of water through the uncapped pit. This rise in groundwater table below the waste pit will cause leaching of the contaminants and the subsequent drawdown will inject the contaminants into the groundwater flow. This phenomenon of supplying of contaminants into the groundwater flow is referred to as mounding. Short heavy precipitation will cause the landfill to leach intermittently. These short pulses or pockets of leachate will spread by diffusion and advection. At the site, this type of contaminant transport is possible due to the nature of the precipitation in the area.

Application of Analytical Groundwater Flow Models

There are a number of groundwater flow models currently in use and among them the most widely used are MODFLOW, PLASM, and AQUIFEM. These models provide approximate numerical solutions to the equations governing the transient flow of groundwater through layered aquifer systems. Because of the ability to model varying material properties, complex geometries, flow loads, and boundary conditions, these computer oriented numerical solutions provide an attractive prediction framework in comparison to analytical solutions, laboratory models, or field observations. At this site, there is not enough field data to fully understand the hydrogeological mechanisms driving the groundwater flow. Thus, numerical modelling provides a valuable tool kit to fill the gaps in our understanding of this site. Because of its reliability and the wide acceptance in the industry, the model "MODFLOW" developed by the U.S.G.S. was used throughout this investigation to determine the characteristics of the groundwater flow at this site.

A model of plan size 3750 ft. by 2500 ft. was used to represent the study area. The aquifer system was composed of a top silt layer of thickness 30 to 20 ft. and a sandy layer of thickness 10 to 20 ft. The third layer also is of thickness 10 ft. and has a permeability equal to that of the top layer. The groundwater level along the northwest boundary was maintained at 863 ft., as shown in field observations in all computer simulations. The high capacity well is located approximately at the center of the study area with the landfills being close to the upper right hand corner of the study area as shown in Figure 3.
The hydraulic flow loads driving the groundwater are as follows:
(a) Pumping at private wells
(b) Recharge due to precipitation
(c) Regional hydrogeology

Results for Groundwater Profiles

Several simulations were performed to determine whether the high capacity well rather than the regional hydrogeology was driving the groundwater flow at this site. The groundwater level north of the landfill was consistently at about 863 ft. in all field

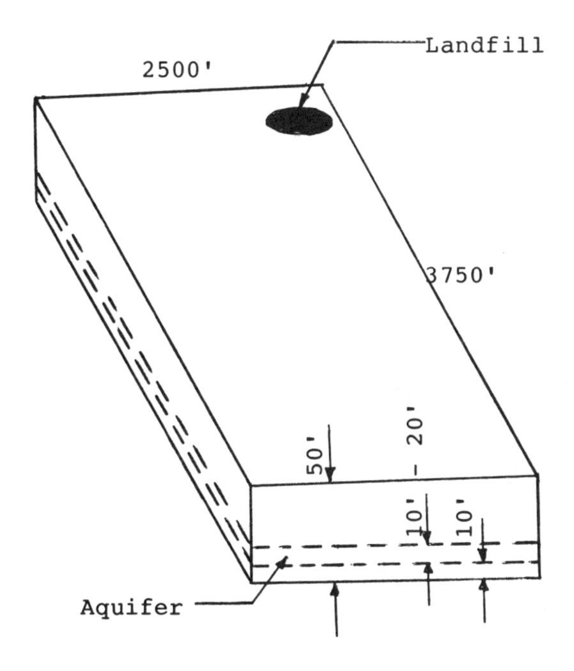

Figure 3. Geometry of the Groundwater Model Used for this Site.

observations. Thus, this value served always as a benchmark for the groundwater surface in the calculations. Twenty-four years of simulation was carried out for the site with all wells pumping at the site. Analyses were also performed for the 24 year period with a recharge of 12 gal/year/sq.ft. over the site, high capacity well at 14.5 Mg/year, and with an aquifer thickness of 10 ft. The results for the end of 24 years are given in Figures 4 and 5. The effects of the high capacity well and the 4000 gal/day well are clearly noticeable in the contours shown in Figure 5.

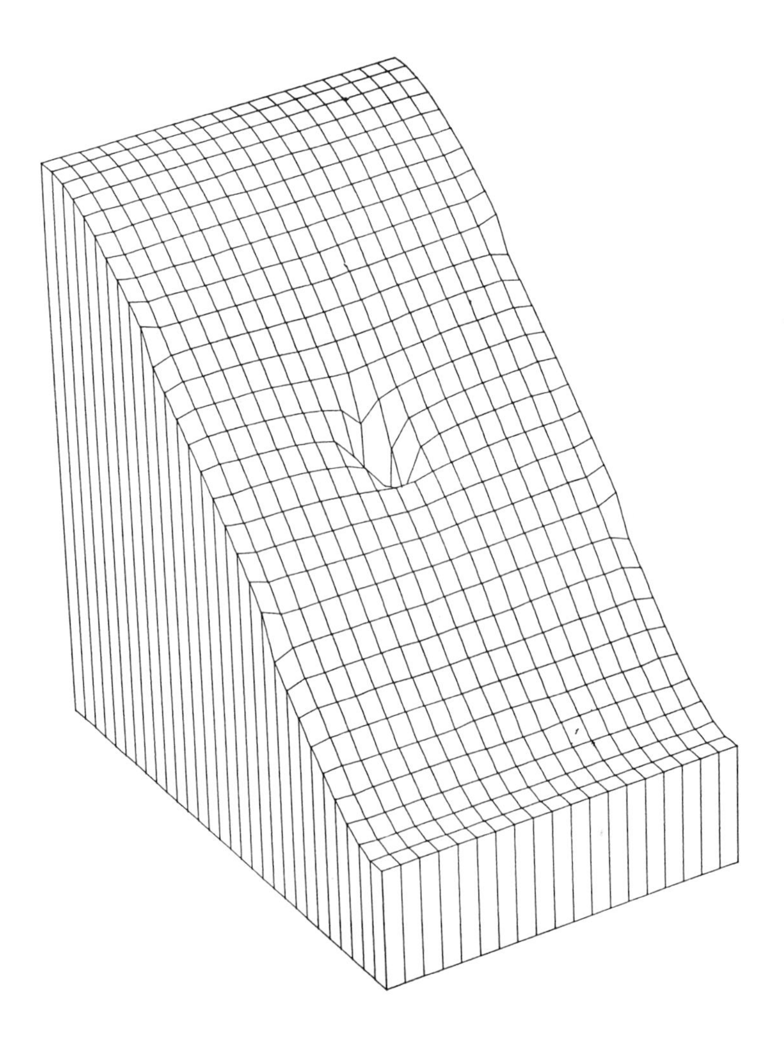

Figure 4. Groundwater Surface By the Model at the end of 24 years

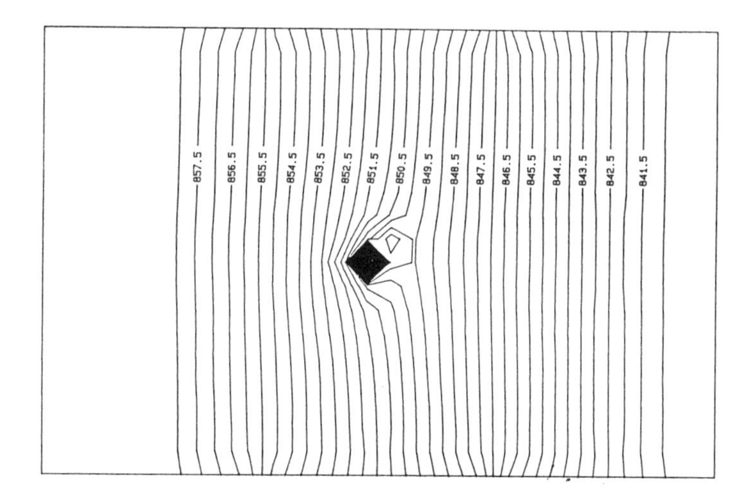

Figure 5. Groundwater Contours at the end of 24 years

Effects of Permeability

The permeability of the 10 ft. thick aquifer was varied by one order of magnitude on either side of the base permeability of 150 ft./day (0.05 cm/sec) and simulations were performed for the groundwater profiles. The drawdown around the high capacity well is naturally greater when the permeability of the aquifer is as low as 0.005 cm/sec.

Effects of Thickness of Aquifer

Due to the fact that the aquifer at this site varies in thickness from 5 to 30 feet, the effects of its thickness on the predicted groundwater flow was analyzed during this investigation. The aquifer thickness did not appear to influence the results appreciably.

Effects of Pumping

The influence of pumping at the high capacity well on the groundwater conditions was studied by shutting the well off completely for 24 years and these results are shown in Figure 6. The groundwater contours in Figure 5 for full pumping and those in Figure 6 for no high capacity pumping appear to be of similar shapes and intensities indicating that the high capacity well pumping do not influence the groundwater conditions at this site significantly.

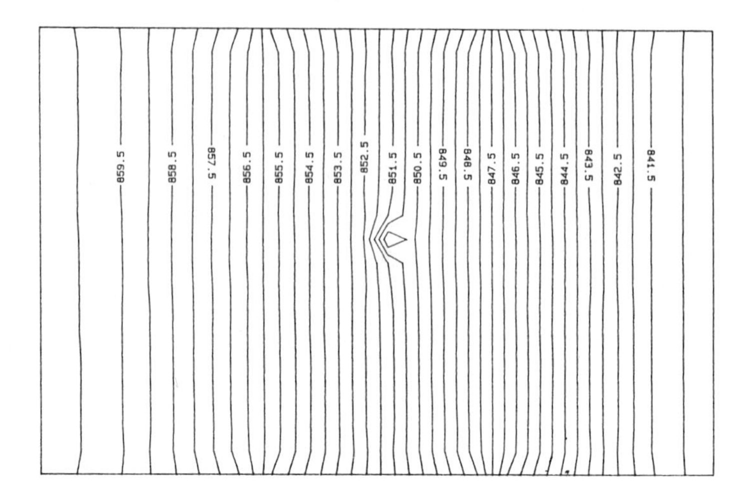

Figure 6. Groundwater Contours
Without High Pumping

Effects of Precipitation

The representative recharge for this site is 12 gal/yr/sq.ft.. In order to study the effects of unusually high recharge of 30 gal/yr/sq.ft. at this site, a simulation was done. The groundwater tends to approach levels higher than those obtained for normal recharge by about 4 ft. Computer simulations for no recharge gave water levels which are about 3 ft. lower than those obtained with normal recharge.

DESCRIPTION OF CONTAMINANT MOVEMENT

Application of Contaminant Transport Models

There are a number of analytical and numerical models for studying the contaminant transport under various flow conditions. The primary mechanisms driving the contaminant movement are as follows:

(a) Advection
(b) Dispersion
(c) Adsorption
(d) Decay
(e) Chemical and Biological Transformation

The primary components of the above physical processes controlling the flow of organics such as those in a fairly previous aquifer system are advection, dispersion, and adsorption. Among the several models used in the industry, Wilson-Miller and Randomwalk models are recommended by most engineers involved in modelling. However, these models are rather complex to use and are not available in well-documented form for three-dimensional applications. Analytical models based on closed form solutions are much easier to apply for three-dimensional transport problems. These models however, have certain limitations. Nevertheless, they are capable of yielding reasonable answers to contaminant transport problems when applied properly and these simple models were used in this investigation.

Behavior of Various Contaminants

The primary chemicals dumped at the landfill and their daughter compounds formed after interaction with bacteria and other organisms found in the geological media were studied for their densities, solubilities, molecular weights, and formulae. The density of many of these compounds is lower than that of water and therefore, these contaminants will float on top of the groundwater surface if they are neither soluble nor miscible in water. These compounds will travel near the groundwater table in the direction of groundwater flow. There are other compounds which are soluable or miscible and these will be retarded in the flow due to adsorption effects in the soil mass. The retardation factor which scales the groundwater advection velocity to give the contaminant transport velocity increases with the organic carbon content of the geological medium and decreases with increasing solubility of the compound in water. The compounds which do not dissolve in water and are heavier than water will sink to the bedrock surface, and will travel along the interface following the contours of the interface. Compounds such as 1,2 Dichloroethylene, Ethylbenzene, P-Dichlorobenzene, and Tetrachloroethylene will take considerably more time than other compounds to reach locations far from the source at this site. The other chemicals such as acetone, and tetrahydrofuron found at this site will travel at about the same or at a lower velocity than that of water.

Material Properties

The permeability of the aquifer was set at 150 ft./day (0.05 cm/sec) and the dispersivity values used are 50 ft., 10 ft., and 5ft. in the longitudinal, transverse, and vertical directions. The decay factor was assumed to be zero. The flow velocity used in most predictions was 0.85 ft/day or 310 ft/yr based on the groundwater flow computations.

Contaminant Loads

Two types of loads were tried for contaminants with two values for the retardation factor, 1.0 and 2.0. The first contaminant load was of strength 1.0 uniformly over the period of 2 or 10 years and the second load was of slug type of intensity 0.2 and 1.0 for 2 year and 10 year durations, respectively. Thus, the concentration contour plots can be scaled by the source strength to arrive at concentration distributions along the horizontal planes at each of the mid-depths of the sublayers forming the aquifer system.

Results for Concentrations of Contaminants

The concentration plots obtained for the end of 2 years at an advective velocity of 310 ft/year for the mid depth of the aquifer are shown in Figure 7. The advective velocity was chosen to be 155 ft/year to approximate the effects of the retardation factor of 2.0 and runs were made for 5 years and 20 years with continuous supply of contaminants at the source. Separate simulations were carried out to determine the concentrations of contaminants when the contaminants enter the groundwater system as leachate pockets. The results for unretarded contaminants for 10 years are shown in Figure 8.

CONCLUSIONS

The conclusions that can be drawn from this investigation are as follows:
(a) The data from the borings at the site indicate that the site is quite pervious. Furthermore, the previous hydrogeologic investigations did not perform permeability tests in the most pervious layers of soil at site. Thus, the more representative value for the permeability of the site is like 0.05 cm/sec.
(b) The data collected from various sources indicate that there was sufficient information available before 1960 in the public domain about the pervious nature of this site.
(c)The detection limits of the instruments used by various analytical labs were higher thatn those recommended by Wisconsin DNR and the U.S. EPA. Therefore, all data reported to be BDL do not preclude the possibility of noncompliance with the State and Federal requirements of groundwater quality standards.
(d)The analyses performed on the local and regional hydrogeology when combined with available field observations indicate that the heavy pumping at the high capacity well do not provide the primary mechanism for the groundwater flow.
(e)The contaminant transport computations, the estimates of various chemicals dumped in the pit, and field observations indicate that the private wells are affected by the dumping of the chemicals and will continue to be affected due to the persistant nature of the organics present at the site.

ACKNOWLEDGEMENTS
The following graduate research assistants were of invaluable help during this investigation: Messrs. James B. Hutchison and Jonathan Voichick.

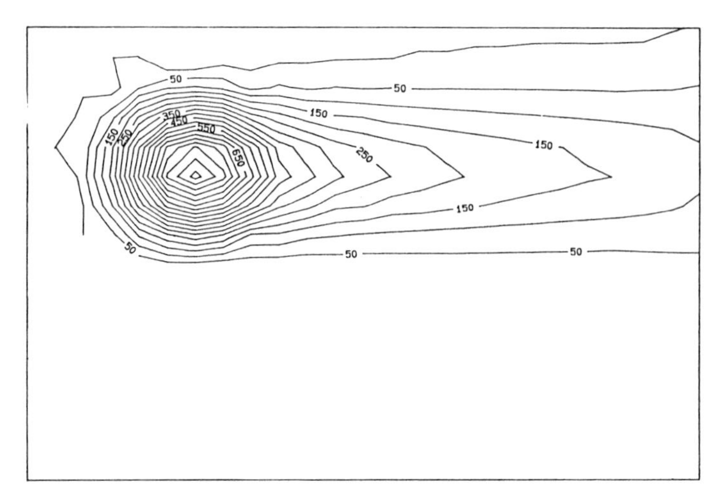

Figure 7. Contours of Contamination Level With Uniform Discharge of Chemicals at the Landfill Site for 2 years

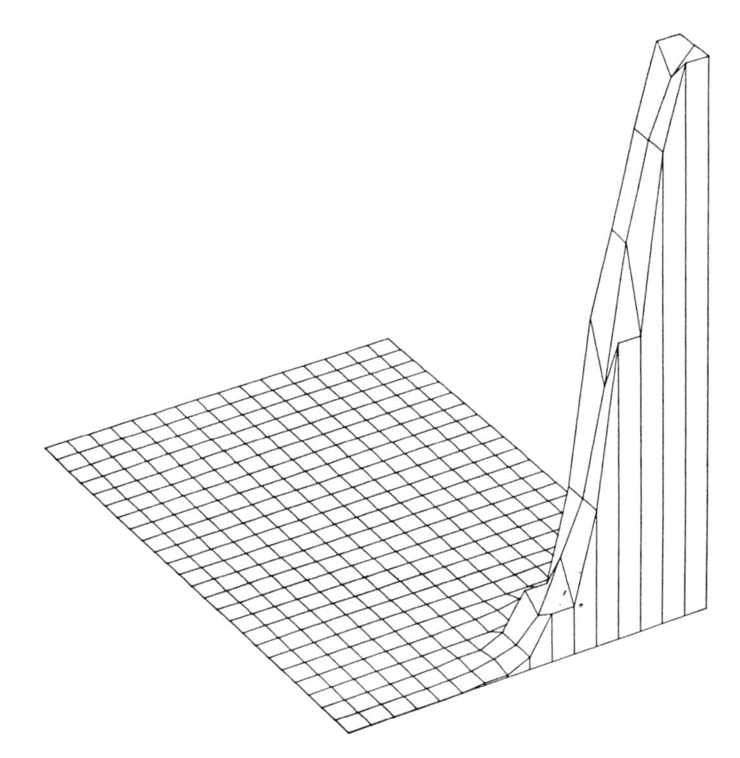

Figure 8. Contours of Contamination Level With Slug Type Discharge at the Landfill Site for 10 years

STABILIZATION OF PETROLEUM REFINING WASTES WITH POZZOLANS

Joseph P. Martin, Frank Gontowski and Lewis T. Donofrio, Jr.,
Drexel University, Philadelphia, Pennsylvania

ABSTRACT

Cement stabilization of wastes from process industries is a growing practice which can limit secondary pollutant releases during operational and post-closure periods of a landfill project. To provide for reliable disposal of waste residuals in a landfill, the properties of the stabilized material must be established in terms of mechanical or dimensional stability, contaminant immobilization and restriction of fluid movement. The in-situ performance of a stabilized waste depends upon development of a mixture and a construction technique appropriate to the stress, climatic and biochemical environment within landfill.

With this philosophy as a guide, a research program was initiated to stabilize petroleum refining wastes with a mixture of lime, fly ash, and in some cases, portland cement. Addition of measured amounts of these materials to the acid hydrocarbon sludge and the spent clay results in a soil-like substance which can be placed and compacted in monolithic cells on the site. To optimize the proportioning for volumetric economy and performance, a series of relatively simple indicators of mechanical, fixation and hydraulic behavior were used. The concentration herein is on the mechanical features of two alternatives, a sludge-clay-pozzolan mixture and a "conventionally" solidified sludge.

INTRODUCTION

Industrial waste stabilization encompasses many practices with the common goal of decreasing the potential for uncontrolled release of contaminants from waste materials in transit, storage or final disposal. Techniques range from reducing bulk mobility, e.g., dewatering, to immobilization of individual contaminant

Geotechnical and Geohydrological Aspects of Waste Management, D. J. A. van Zyl et al., Eds., © 1987 Lewis Publishers, Inc., Chelsea, Michigan—Printed in USA.

species in insoluble matrices [1]. For industrial waste disposal in landfills, one of the most popular methods is mechanical stabilization with hydraulic cements [2]. While this method is primarily used to solidify sludges, cementing particulate wastes into a continuous mass is also done [3]. When large volumes of raw waste must be landfilled, low-cost industrial byproducts such as fly ash or kiln dust are frequently used in pozzolanic cementing systems [2,4,5].

Developments in landfill technology have tended to proceed in two tracks, with one concentrating on external containment and the other on stabilizing the contents of an enclosure. In the former approach, the goal is prevention of liquid entry (cover), and interception and diversion of liquid pollutants accumulating at the base (liner-leachate collection). In contrast, stabilization ultimately focuses on restricting internal generation and movement of contaminants in liquid, gas and particulate forms. Provision of dimensional stability to the final cover and hampering permeation of infiltrated liquid are but two of the symbiotic effects of combined containment and stabilization. When technically feasible, using elements from each practice provides a belt-and-suspenders approach to minimizing the risk of impacts on the surrounding environment [4].

Waste stabilization has the highest potential at captive sites containing wastes from bulk process industries. In such well-characterized and controllable situations, opportunities often exist to optimize a waste disposal solution. However, much development of stabilization technology has occured on a proprietary basis, with limited documentation of construction feasibility and properties of the resulting material. There is also a natural tendency towards one-size-fits-all solutions. The true test of stabilization performance is the response of the fill to the in-situ stress, climate and biochemical environment as illustrated on Figure 1, so that extensive site specific investigation prior to construction is desirable to enhance performance.

This paper describes a study to stabilize petroleum refining wastes with a pozzolanic mixture of hydrated lime and fly ash, supplemented in some cases with portland cement. The wastes include about one million cubic yards of acid hydrocarbon sludge and spent fuller's earth (attapulgite) clay, deposited in a series of piles and lagoons. The anticipated remediation plan involves constructing a series of onsite landfills by stabilizing each deposit in turn. Two basic techniques were studied, one involving a conventional admixture of cementing material to the sludge to form a microencapsulating solid skeleton, and the other using the clay as the aggregate which is cemented into a stable porous matrix to entrap the sludge. The site capacity limits admixture volume, such that optimization of proportioning to provide the best values of an array of desired properties was the focus of the empirical investigation. With the step-by-step study, a series of indicator properties were used to guide project development.

BACKGROUND

Design to reduce the potential for internal pollutant generation and release internally requires identification of the mechanisms of contaminant mobilization and movement [6,7]. For contaminants deposited in a landfill to reach the boundary, several requirements exist:

- The contaminant must be in mobile form.
- The contaminant must have access to an internal pathway.
- A gradient must exist to induce transport.

Deposited wastes may contain free liquid or soluble constituents, while biological or chemical reactions may form byproducts more mobile than the parent materials. Infiltration of precipitation and groundwater into the landfill is one source of leachate, dissolving contaminants or displacing immiscible fluids. However, during and shortly after filling, excess liquid will drain by gravity, and interstitial liquids can be expelled during compression as overburden loads are placed. If a continuous water column exists in the fill pore spaces, solute transport by diffusion may also be possible with no bulk liquid movement. Each form of leachate may have different contaminant species and concentrations.

Noxious gases may also be originally present in the waste, and more may be produced by reactions and volatilization at gas-liquid interfaces [8]. Gas movement can be driven by diffusion, displacement by infiltrating liquid, consolidation expulsion, and barometric pumping. Particulates can be conveyed by wind, erosion, and rodents.

In terms of the secondary pollutant types and risk of generation a landfill project can be divided into two phases: operational or filling and post-closure. During the filling phase, the liner-leachate collection system is the primary defense against liquid escape from the enclosure, but does little to mitigate airborne releases of noxious gases or particulates. All incident precipation is expected to produce leachate by percolation through or flow over the exposed waste. Daily cover provides some control of biological vectors (rodents, etc.) and wind-blown particulates, but expensive space in the enclosure is diverted by this effort. At closure, placement of the final cover reduces most of these problems, but not permanently if dimensional stability (alignment and structural support) is not provided by the waste on which the cover rests.

CEMENT STABILIZATION EFFECTS

With this background, it can be seen that stabilization should mitigate problems inherent in exposed deposition of a pervious or mobile fill and not resolved by subsequent isolation with an external containment envelope.

A minimum level of waste pretreatment is removal of free liquid content by prior drainage, solidification or in-situ slurry dewatering through the leachate collection system. The popularity of hydraulic cementing is due in part to the uptake of liquid in hardening, but sufficient moisture must be present to sustain hydration [6]. Further improvements with some degree of stabilization can be classified into three types of effects: mechanical stabilization, contaminant fixation or immobilization, and restriction of internal fluid transport.

Mechanical stabilization can either involve hardening wastes on a small scale, as in drum solidification, or forming the whole deposit into a continuous mass, as shown in Figure 1. If a fairly impervious monolith results, contact between the waste and the local environment is theoretically restricted to the boundaries which can then be sealed to the appropriate degree with an external containment.

Whether a mechanically stabilized deposit is impervious or not, the final landfill cover will have better dimensional stability. A cover is a roof like any other in that its performance depends as much on alignment and structural integrity as on the materials from which it is made. Improving the shear strength and stiffness of the deposit reduces the need for camber in the closure slopes. With an unstabilized fill, it may be necessary to place cover slopes at high angles to compensate for post-closure settlement [9], but this raises concern about creep distortion, accelerated erosion, and slope failure.

During the operational phase, cement stabilization may improve construction efficiency and reduce environment impacts. Hardened fill supports the machinery placing new lifts. Stiffening of lower layers decreases gas and liquid expulsion by consolidation, especially if the rate of filling is chosen with consideration of the hardening rate. A cemented waste is not necessarily impervious, but would have a higher runoff coefficient, be less erodible, and be easier to grade for rapid runoff than loose fill. These factors limit the contact between incident precipitation and the waste, and improve collected leachate quality from a working site.

While mechanical stabilization is intended to improve large scale properties, immobilization or fixation attempts to restrain individual contaminant molecules or droplets at their deposited locations. The most desireable situation is fixation in stable lattices by irreversible reactions [1], but reducing solubility with pH or other adjustments may have the same effect as long as reversal is prevented. Cement hydration incorporates some inorganics in cement precipitates, and raises the pH as well. However, many contaminants, especially organics, cannot be chemically fixated to the degree that may be desired. Another form of fixation is coating or entrapping droplets and grains, physical isolating them by microencapsulation in a porous structure. This is the primary form of immobilization attributed to cement stabilization [1].

If mobilized, contaminants need a pathway and a fluid carrier to continue transport. Fluid transport is generally described by either Darcy's or Fick's laws, describing flux of liquid, bulk gas, solute or gas species in terms of both a property of the matrix and a potential gradient. Saturated water permeability is an indicator of the resistance of the matrix to both leachate flow and solute diffusion, as it depends upon factors such as porosity, pore size distribution, tortuousity and liquid viscosity. Desaturation reduces liquid permeability, but increases source volatilization and diffusion or bulk flow of gases. Fluid transport gradients in terms of total head, gas pressure or concentration can only be evaluated in terms of the layout and boundary conditions of a specific site. The gradients also vary between operational (exposed, compressing and draining) and post-closure periods.

The bulk integrity of a monolith in its field environment is as important as the permeability of intact sections. Cracking opens internal pathways and weathering or reactions can damage a microencapsulating structure. Important material properties are freeze-thaw durability, tensile strength, shrink-swell potential, etc. Layout and subgrade preparation are also important. Cracking from flexural stresses is of most concern in the upper and lower parts of a fill, and is dependent on slope angles and the uniformity of subgrade support. Only the upper parts of a fill are subject to shrinkage or freezing, such that poor behavior of a stabilized waste in these areas can be compensated with insulating cover design or special waste mixtures.

CAPTIVE INDUSTRIAL WASTE DEPOSITS

Landfills can be characterized in a number of ways, but one critical distinction is on the basis of homogeneity and multiplicity of waste sources. Captive disposal sites for process industries are one important subclass. Continued exposed storage of residuals that cannot be recycled, incinerated, etc. in waste piles and sludge lagoons may present unacceptable maintenance costs or offsite environmental effects. Execution of stabilization might be complicated by one or more of the following conditions:

- Large waste volume, often from years of accumulation. Unit costs and construction efficiency are critical factors.
- The basic nature of the wastes are known, but the character varies from point to point due to weathering and process changes.
- The site is cramped, presenting logistics problems and restrictions on stabilizing additive volume.
- Short-term offsite impacts control remediation activities.
- The site is a prime economic location, prompting a desire to reclaim the site for re-use.

When some or all of these factors apply, there is a need for sensitivity studies to obtain the most effective solution and to provide a data base to allow efficient response to variations in field conditions.

PROJECT DESCRIPTION

All of the factors listed above apply to the petroleum refining waste disposal site, where acid hydrocarbon sludge is stored in several pits or lagoons and the spent clay is distributed between several piles. The clay originally contained oil adsorbed in product upgrading, but hydration has casued swelling and hydrocarbon expulsion. Groundwater interception and lagoon decanting systems are in place at the site, but these measures are considered temporary. A system of onsite stabilized landfills is thought to be among the least risky alternatives from both technical and political viewpoints. Index properties of the waste materials are shown on Table 1.

Table 1. Waste Material Properties.

Acid hydrocarbon sludge	
Specific gravity:	0.95 - 1.2
Volatiles:	40% - 80%
Water Content:	10% - 30%
Sulfur:	2% - 10%
Ash:	10% - 50%
pH:	2.0 - 2.5
Spent fuller's earth	
Predominant clay mineral:	Attapulgite
Specific Gravity:	1.98
Liquid limit:	192%
Plastic limit:	23%
In-situ moisture content:	80% -100%
Hydrocarbon content:	6% -10%
Standard Proctor dry density:	52 lb/ft^3

It is difficult to measure the hydrocarbon content of the clay or a candidate stabilized mixture on a routine basis; therefore, all moisture contents noted hereafter include both water and volatiles loss after oven drying for 24 hours at 105°C. Another waste under consideration for co-disposal, but not presently on the site, is spent catalyst fines. These are dry, hardened clay minerals in the form of silt-sized spheres with and average diameter of 0.1 mm.

There are two basic stabilization alternatives, one being separate handling of the sludge and clay, and the other a co-disposal scheme. The former involves regrading the clay piles, possibly with reduction in the platicity of the clay, and conventional solidification of the sludge with dry additives that provide an artificial solid skeleton. Co-disposal involves use of the clay as the main aggregate of a cemented solid solid skeleton. The acidic sludge must be neutralized in either case, but neutralization also separates the emulsion, giving the opportunity to entrap or encapsulate individual droplets. The clay was used as an adsorbent in the first place, such that a clay-based solidification involves positive re-use of that waste material.

Major concerns include volumes, constructability and quality control. The site is quite crowded, and to be able to sequentially stabilize lagoons and waste piles as individual cells, the stabilized volume is limited to 50% above that of the existing waste.

In-situ injection of dry cementing materials into the sludge deposits might be feasible, but this technique has obvious problems of quality control. Extra volumes of dry materials would have to be added to assure that no free liquid pockets remained. Therefore, the favored technique is to mix wastes and additives aboveground on a bed or in a mill to obtain a soil-like material that can be compacted to harden in place. This would yield a uniform final product with both measureable mechanical properties and economical use of additives. However, the concerns about air quality effects during construction require that in-situ mixing continue to be considered as an alternative pending results of comphrehensive atmospheric monitoring during a pilot field study.

Candidate additives include portland cement, lime and fly ash. Dry additives are needed simply to make a fresh soil-like mixture. Fly ash is available from a nearby powerplant, but it was found to be minimally self-cementing when mixed with water. Lime is required for neutralization, and initiates the pozzolanic reaction. It also conditions the clay for environmental (shrink-swell) stability and easier construction handling. A blend of lime and fly ash gives good results in stabilizing a mixture of sludge and clay, but replacement of some of the lime with portland cement was required to solidify the sludge. Properties of the additives are shown on Table 2:

Table 2. Additive Properties

Portland Cement	Type II
Lime	
Type:	Hydrated, Air Entraining
CaO:	42.5%
MgO:	29.2%
H_2O:	24.2% (combined)
Passing #30 mesh:	99.8%
Passing #200 mesh:	87.1%
Fly Ash	
Source:	Bituminous coal combustion
Specific gravity:	2.52
d_{50} (mm)	0.3
C_u (coef. of uniformity)	8.8

Other characterization studies are underway to determine the free lime and glass content of the fly ash, to allow rapid prediction of the efficiency of ash from other sources if needed in final remediation of the site.

LABORATORY STUDY GOALS AND TESTS

With the large volumes and complex reactions between several components, an extensive sensitivity study was initiated to optimize proportioning and procedures and to evaluate the properties of stabilized mixtures. Construction workability dominated the initial work. The first goal was identification of fresh mixtures that would best use onsite materials, have no free liquid, be workable with conventional construction machinery, and have minimum volumetric swell. Fairly rapid strength gain after compaction is also desirable to support equipment and to mitigate exposed operational period impacts as discussed above.

For permanent dimensional stability, it was desired that the hardened material attain the strength and stiffness of stiff clay. Desired results in regards to fixation and fluid transport control aspects of stabilization included non-acidic leachate, minimal mobilization of hydrocarbons in permeation, and low permeability. It was clear that a rigid set of numerical results could not be obtained with any one proportioning, so that an array of different mixes emphasizing one aspect or another were anticipated.

A sequence of indicator properties were evaluated with each promising mixture:

- Unconfined compressive strength.
- Ductility indexes derived from the stress-strain curves.
- One-dimensional compressibility.
- Split-cylinder tensile strength.
- Falling-head permeability.
- pH of effluent (leachate) from permeability tests.
- Clarity of permeability test effluent, a rapid indicator of hydrocarbon dissolution or displacement.
- Rate of evaporation from compacted Proctor molds.
- Shrinkage cracking in the drying mold samples.

INITIAL STEPS

Before substantial full-mix trials were run, individual component reactions were evaluated. To indicate optimum conditions for cementing, mixtures of fly ash and lime in ratios from 3:1 to 5:1 were tested, with the expected result of high strength in rich mixtures. Optimum moisture contents in all cases were between 45% and 50%, indicating a high potential for free liquid uptake in sludge stabilization.

At the natural water content, the clay could not be disaggregated , but this was possible when the clay was dried to about 60% moisture content. Further drying is unwise, as this material showed high swell propensity in one-dimensional free-swell tests. Rehydration and accompanying swelling of the clay resulted in expulsion of the little oil remaining after years of exposure, obviously reducing sludge entrapment value as well. Consequently, some of the added lime was "alloted" to reducing clay plasticity and water affinity. Both water and lime are essential components of pozzolanic cementing, and much of the empirical study revolved around proportioning and mixing order variations to satisfy competing demands for water and cations by the clay and the pozzolanic reaction.

Sample preparation also required development and standardization. To mimic the simplest, but not necessarily the most efficient, field procedure, all mixes were blended and mixed by hand in tubs. Determination that a particular mixture was of a soil-like texture was a fairly subjective estimate, but an unconfined compressive strength of 5 lb/in^2 with a fresh mixture compacted at 80% of Proctor effort was used as a confirming index. After several trial methods, the technique settled upon to produce such specimens was compaction with a Harvard miniature compactor in 1.5" or larger pure teflon molds. Specimens were extracted immediately after compaction for curing in a freestanding position. Other vessels, such as permeameters and consolidometers were prepared in a similar fashion.

While strength gain and other parameters were measured with specimens which were cured in a dry, ventilated atmosphere or in a conventional moist cabinet for concrete curing, most results were obtained with samples cured in sealed containers.

Field exposure and confining stress will vary with the rate of construction and the weather. To test both shrinkage and relative desiccation/volatilization rates, samples of varying composition and compaction effort were packed in molds and allowed to vent through the upper surface. Actual rates of clay-based specimen drying were slow enough [6] to assure that sealed curing was a reasonable representation of field moisture conditions in the bulk of a fill under construction.

An accounting system was needed to identify each mixture, and the most convenient procedure for the laboratory phase was to use a weight proportioning in the order:

sludge/ clay/ fly ash/ lime/ catalyst fines

In the field, proportioning will probably be volumetric. With the higher specific gravities of the additives, volumetric increase from stabilization is less than that indicated by the weight proportionings discussed below.

SLUDGE-CLAY POZZOLAN MIXTURE RESULTS

All mixtures including both sludge and clay were prepared with equal wet weights of each, as this is roughly the current onsite ratio. The clay was dried to a 60% moisture content and in some cases, pre-stabilized with an admixture of lime. Tub tests showed about 0.75 to 1.0 parts fly ash was needed to adsorb sludge moisture and form soil peds when no lime was included, i.e., 1.0/1.0/0.75 to 1.0/ 0/0 by weight.

The first suite of tests was run with equal weights of sludge, moist clay and fly ash, which actually represents a volumetric ratio of 1: 0.75 : 0.48, respectively. Lime content was varied from zero to 0.5 parts, by weight. Results of clarity tests (percent light transmittance at 660 x 10^{-6}m) on permeation effluent are shown on Figure 2. This will later be correlated to the chemical oxygen demand (COD) of the leachate. The lowest curve on Figure 2 is for the zero lime sample, indicating severe hydrocarbon mobilization. A pH of 7 was obtained, no immiscible fluids were displaced, and organic content was in dissolved form only. This indicated basic success in neutralization and hydrocarbon separation with the minimal amount of free lime in the fly ash. An unsteady permeability with continued throughput was also observed, initially at 1.0 x 10^{-7}cm/sec, but peaking at 5.0 x 10^{-7} cm/sec at five pore volumes.

The curve for the high lime (0.5 parts) permeation test shows very high clarity, but a pH of 12 and permeability in the 10^{-5}cm/sec range also resulted. Neither extreme in lime content showed good mechanical properties, the zero lime sample having poor bonding and the high lime showing excess, friable and soluble powder. Therefore, the results shown on figure 2 favored further study of mixtures with lime contents in the 0.2 to 0.25 parts range.

Figure 3 shows results of unconfined compression of with this general range of lime content. The high fly ash specimens (1.0/1.0/1.0/0.2/0) showed finite but limited strength gain (lowest curve). Reasonable stiffness, was observed in one-dimensional compression, as illustrated with the middle curve of Figure 4. Addition of large proportions of spent catalyst fines liquified fresh samples, and even small amounts (0.2 parts) significantly increased compressibility, as shown by the lowest curve of Figure 4. The conclusion from all of these tests was that the clay adsorbed much of the water and cations, leaving poor cementing conditions and excess unreacted fly ash. The spent catalyst fines did not appear to take part in the pozzolanic reaction at all.

To increase strength without affecting effluent pH, one alternative was to decrease the fly ash : lime ratio, thus making better use of the ash and reducing cementing liquid needs i.e. 1.0/1.0/0.75/0.2/0. Figure 3 (middle curve) shows significant strength improvement, Figure 4 shows mildly increased stiffness with reduced fly ash, and Figure 2 shows that the hydrocarbon dissolution was not affected. Effluent pH values in the range of 8.0 to 8.5 were obtained with lower fly ash, along with a stable permeability on the order of 6.0 x 10^{-7}cm/sec.

A re-check of lime sensitivity with the reduced ash content still showed the best strength results with 0.2 to 0.3 parts lime, or 6% to 8% of the total mixture. This also yielded intermediate values of ductility, indexed for this study by the strain at sample failure. Insufficient ductility would present a risk of brittle cracking, whereas high ductility might cause slope creep problems. The volumetric proportioning of the reduced fly ash mixture is 1.0/0.75/0.36/0.085, roughly 45% sludge and 34% clay overall, or a 26% swell over the original waste volume.

If convenient for construction purposes, stockpiled moist clay can be pre-stabilized with lime. It was found that this also improves most properties of interest. The top curve on Figure 3 shows that admixture of 5% lime to the clay before blending with the other materials significantly increases strength. The premix lime content is accounted for with a parenthesis in the "active" ingredient slot, i.e.,

1.0/1.0/0.75/0.2(0.05)/0. This variation is the focus of the current suite of tests, which also show favorable results in freeze-thaw durability and permeability. Consequently, a pre-stabilized clay mix could be used as a base course and also as the top layer in a reclaimed pit, with the balance of the backfill being less complex proportionings.

SLUDGE-POZZOLAN MIX RESULTS

While the cemented clay forms an artificial skeleton (10) of the mixes just described, offsite materials are required for this purpose in the non-clay sludge-pozzolan mixtures. A major concern is adaptability to different sludge consistencies. With the clay mixes, similar results were obtained with a fairly wide range of sludge consistencies, such that it could be said that the clay "dampens" the effects of variations in sludge viscosity. However, to provide a workable fresh mixture with the non-clay mixtures, more lime-fly ash dry admixture was needed as the sludge became "soupier". The results shown herein are for mixtures using a sludge location that contains some solids. A soil-like texture was obtained with a sludge to dry additive weight ratio of 1.0 : 0.8, which actually includes 75% waste by volume, a very efficient proportioning if hardened properties were suitable.

The basic dry blend is a three to one ratio of fly ash to "active" ingredients, using lime as the reference. Figure 5 shows stress-strain curves at seven days curing for different levels of lime replacement with portland cement. The bottom curve is for a sludge-pozzolan mixture only, i.e., 1.0/0/0.6/0.2/0, with no cement. It can be seen that this material is relatively soft. To increase strength gain, some of the lime was replaced with cement. The formal accounting notation is to place the cement proportion in the "active" slot ahead of the lime, i.e., 25% replacement of lime by cement is 1.0/0/0.6/(.05) 0.15/0 , but the shorthand notation is 2.9% cement, whereas the 5.5% cement curve represents a mix with equal cement and lime portions. The fourth curve on Figure 5 is for a slightly less viscous sludge which required equal weights of sludge and dry material, with an overall cement portion of 3.2%.

Adding small amounts of cement increased seven-day strength, as can be seen on Figure 5, but at the full 30 day curing, however, the strengths of the 0%, 2.9% and 3.2% cement mixtures were nearly identical, between 13 and 16 psi. To increase fully cured strength above that obtained with lime and fly ash only, it appears that at least 5% cement is required. The 5.5% cement specimens reached 27 psi unconfined compressive strength in a month.

Addition of cement also produced lower permeability. The zero cement mix had a permeability of 3 x 10^{-6}cm/sec, while the 5.5% cement mix samples showed values in the 10^{-7} to 10^{-8} cm/sec range. With several clay and non-clay mixtures showing permeabilities in that order, self-lining with stabilized waste is technically feasible.

The last plot, Figure 6, shows the importance of curing time on the compressibility of both clay and non-clay mixtures. The two curves showing the most deformation represent the sludge-pozzolan mixtures with no cement, one loaded immediately (lowest curve) and the other after seven days of sealed curing (open squares curve). Both samples showed some collapse upon innundation at low overburden pressures. The other two curves are clay mixtures similar to the heavy fly ash mixture shown on Figure 4. The uppermost curve represents loading after one week of curing, with wetting at approximately 5 feet of overburden pressure. The fourth curve represents the same mixture loaded immediately after packing. The lack of bond formation time resulted in a softer, but not soaking-sensitive behavior. The construction rate and cell depth should therefore be considered in determining mixture proportioning; rapid construction requires faster strength gain. Sludge-pozzolan mixtures without clay or cement are the least complex blends, but should be used only for shallow, slowly constructed cells.

CONCLUSION

A laboratory program to optimize construction and in-situ performance factors for a complex industrial waste stabilization project has been described. The goals of stabilization were divided into mechanical, fixation and fluid restriction elements, and a series of indicator properties evaluated at each stage of an empirical study. Two types of pozzolanic stabilization were investigated, one using spent clay as the microencapsulation skeleton, and the other employing conventional cement-pozzolan solidification of a sludge. Pending the outcome of a pilot field study, it is concluded that either system is feasible, with particular mixtures to be used in different locations within the site which vary in terms of mechanical stress and environmental setting.

REFERENCES

1. Malone, P.G., and Larson, R.J. "Scientific Basis for Hazardous Waste Immobilization", Hazardous and Industrial Solid Waste Testing: Second Symp., ASTM STP 805, R.A.Conway and W.P.Gulledge, eds., pp168-177, (1983)
2. U.S. EPA/U.S. Army Waterways Exp. Sta. "Survey of Solidificatiuon/Stabilization Technology for Hazardous and Industrial Wastes", EPA-600/2-79-056, 161p, (1979)
3. Manca, P.P., Massacci, G., Massidda, L. and Rossi, G. "Stabilization of Mill Tailings with Portland Cement and Fly Ash", Trans. Inst. Mining and Metallurgy, V 93, pp A48-A55, (1984)
4. Smith, C.L. and Frost, D.J. "Secure Landfilling with Pozzolanic Cementing", Proc. 1st Ann Conf. on Hazardous Waste Mgmt., Phila, Pa., pp153-160, (1983)
5. Morgan, D.S., Novoa, J.I. and Halff, A.H. "Oil Sludge Solidification using Cement Kiln Dust", J. Env. Eng. Div., ASCE

110 (EE5), pp 935-949, (1984)
6. Martin, J.P., Koerner, R.M., Felser, A.J. and Davis, K.J. "Load Bearing Properties and Durability of Stabilized Waste", Proc. 38th Canad. Geotech. Conf., Edmonton, Alta., pp 375-380 (1985)
7. Martin, J.P., Van Keuren, E., Raymond, A.J. and Felser, A.J. "Stabilization of Petroleum Refining Wastes for an Onsite Landfill", submitted to ASCE Specialty Conf. for Waste Disposal, Ann Arbor,MI, June 1987
8. Thibodeaux, L.J., "Estimating the Rate of Air Emissions from Hazardous Waste Landfills", J. Hazardous Materials, V 4, pp 235-244, (1981)
9. Yen, B.C., and Scanlon, S., "Sanitary Landfill Settlement Rates", J. Geotech. Eng. Div., ASCE 101 (GT5), pp 475-487 (1975)
10. Winterkorn, H.T., "Soil Stabilization", Foundation Eng. Handbook, H.F. Winterkorn and H.Y. Fang, Eds., Van Nostrand Reinhold (1975)

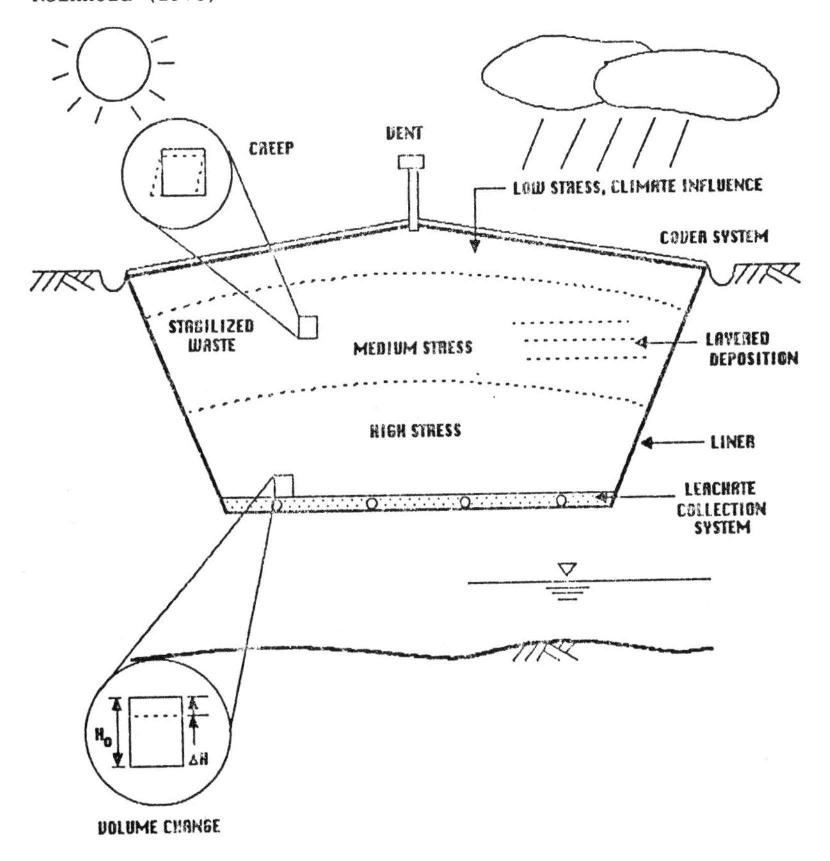

Figure 1. In-Situ Environment of Stabilized Waste

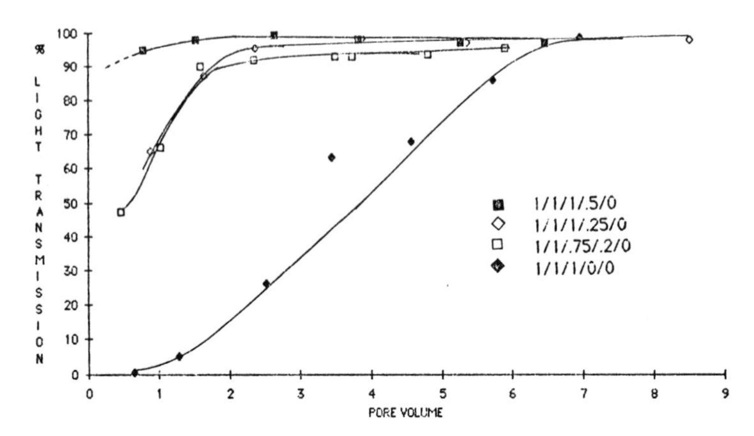

Figure 2. Clarity of Effluent from Permeation of Sludge-Clay Mixtures

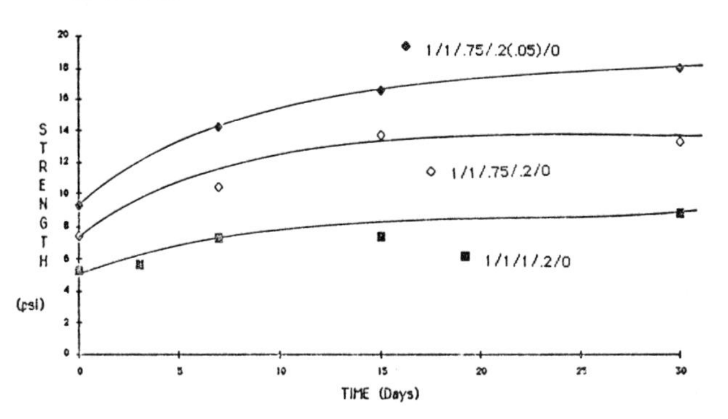

Figure 3. Fly Ash and Clay Pretreatment Effects on Strength Gain

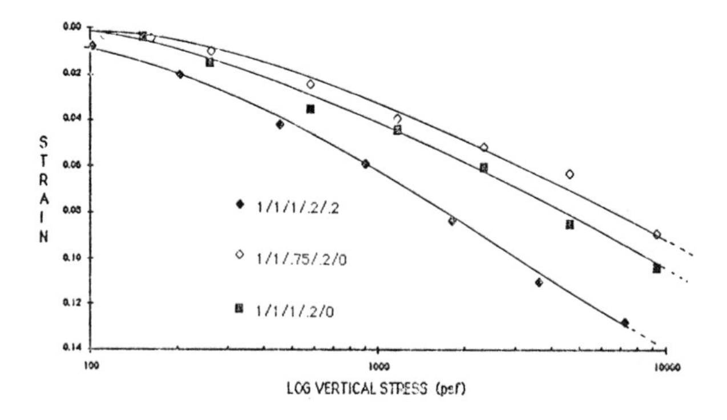

Figure 4. Fly Ash and Catalyst Fines Content Effect on Compressibility

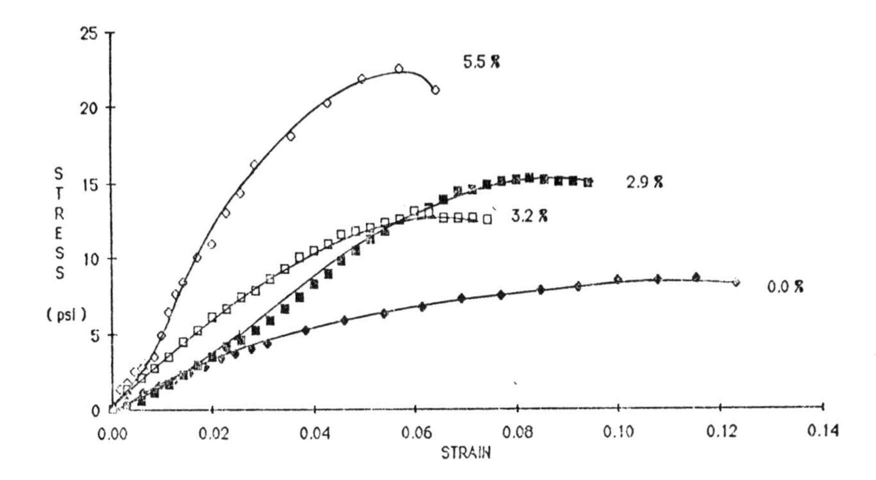

Figure 5. Cement Content Effects on Non-Clay Mixture Strength and Stiffness

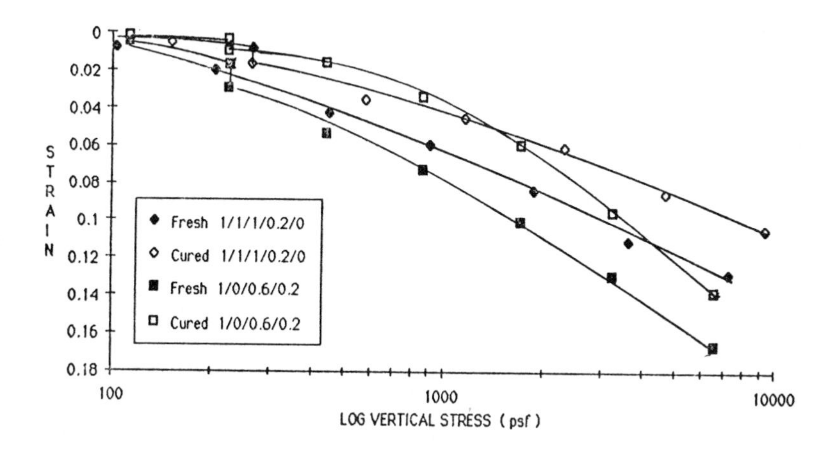

Figure 6. Curing Time Effects on Compressibility of Clay and Non-Clay Mixtures

STABILITY ANALYSIS & DESIGN OF LANDFILLED INDUSTRIAL SLUDGE

Michael J. Monteleone, Geotechnical Engineer,
AWARE Incorporated, West Milford, NJ

John C. Robins, Geotechnical Engineer,
Lippincott Engineering Associates, Delanco, NJ

INTRODUCTION

The stability analysis and design methodology presented within this paper was necessitated by the circumstances present at the site. Since the name and location of site is irrelevent to the problem and analysis, it shall not be revealed.

The situation encountered at this site involved the closure of an existing landfill Cell A, and the construction of a new landfill Cell B, adjacent to the cell to be closed (see figure 1). Along with the usual problems encountered on a job of this nature, the following special problems were of concern:

- Consolidation of the industrial sludge in Cell A due to the surcharge from Cell B.
- Stability of the north slope of Cell A during the construction and operation of Cell B.
- Determination of physical and mechanical properties of an industrial wastewater sludge.

While the literature regarding the analysis required for these special geotechnical/environmental problems is essentially non existent, these problems are very well documented for soil and rock conditions. Therefore, while this type of material is vastly different from that which is typically encountered, the analysis required to investigate a problem of this nature may be found in traditional geotechnical methods. Traditional stability and settlement calculations may be applied to solve these types of problems, which are frequently encountered in landfill design. The following sections will illustrate the use of geotechnical engineering in the design of multi-cell landfill facilities.

Geotechnical and Geohydrological Aspects of Waste Management, D. J. A. van Zyl et al., Eds.,

DESIGN METHODOLOGY

The following design methodology was employed to investigate the aforementioned problems:

- Waste characterization
- Consolidation/settlement analysis
- Side slope stability

The material to be landfilled is a dewatered industrial wastewater treatment plant sludge, with an average solids content of 30%. This industrial sludge was sent to a local engineering/soils lab for determination of the material's physical and mechanical properties. Two types of tests were selected to evaluate the properties of this material. The first test was a standard one dimensional consolidation test. The results of this test provided information as to the consolidation characteristics of the sludge. This data was subsequently utilized in settlement and consolidation calculations. The second test was a consolidated-undrained (C-U) triaxial test. The results of this test provided information as to the material's shear strength, cohesion and angle of internal friction. This data was used in a slope stability analysis which will be presented in the sections to follow.

The results of this testing program are presented in Table 1.

Table 1. Summary of Geotechnical Properties of an Industrial Sludge

<u>Physical</u>

- Unit weight = 30 pcf
- Moisture content = 192%
- Void ratio = 4.42

<u>Mechanical</u>

- Angle of internal friction = 11°
- Cohesion = 403 psf

Armed with this data base, the stability of the sludge and the manner by which this material will impact the overall design of Cell's A and B may now be evaluated.

Two cell configurations were considered. Alternative one positioned Cell A adjacent to Cell B (see figure 1), and alternative two considered a backfilled sand wedge adjacent to Cell A (see figure 2). The problem of concern common to both of these cell configurations was a load distribution to be placed on the north side slope liner of Cell A. This load would place the contents of Cell A industrial sludge under compression. As a result, it became necessary to calculate the amount of consolidation the sludge would undergo, given these two loading conditions. Excess consolidation of the sludge in Cell A could place undo stress on the side slope liner of this cell. Utilizing Terzaghi's Consolidation Theory and the data obtained from the lab investigation, the amount of expected settlement and the time for this settlement to occur was computed. The analysis and results will be presented in the sections to follow.

The second area of concern was the stability of the north slope of Cell A during and after the construction of Cell B. The Simplified Bishop Method of Slope Stability Analysis was employed to determine the stability of this situation. The analysis included an evaluation of the slope, subjected to both static and dynamic loading conditions. The results of this analysis will be presented in the sections to follow.

ANALYSIS AND RESULTS

Settlement Analysis

Due to the nature of the material to be landfilled, a time dependent method of analysis was sought to model the consolidation behavior of the landfilled sludge. Terzaghi's Theory of Time Dependent Consolidation was selected to model the consolidation behavior of the sludge. From the laboratory results presented earlier, it appeared that this material was not elastic by nature but rather plastic, similar to a soft organic clay. Some assumptions were made in order to simplify the analysis. These assumptions tended to favor a "worst case" scenario, thereby yielding conservative results. The following assumptions were used:

- Single drainage path
- Uniform load distribution
  (in reality the load distribution is triangular)
- No load spreading allowed
- No support from the geosynthetic composite system
- No biodegradation of the waste

Table 2 presents the results obtained from this analysis.

Table 2 Summary of Settlement Analyses

Cell Configuration	Total Settlement (in.)	Percent Consolidation (%)	Time to Occur (yr.)
Cell A adjacent to Cell B (Figure 1)	.7	100	10.1
Cell A adjacent to sandwedge (Figure 2)	2.7	100	10.1

Slope Stability Analysis

The last concern with this multicell facility was the stability of the north slope of Cell A. Cell A was to be closed before construction of Cell B was to proceed. Therefore, the side slope of Cell A would be a free standing slope with no support, while being subjected to static and dynamic loads.

Due to the large number of methods available, and the time constraints placed on the project, only one method was employed to analyze this problem. The method selected was the Simplified Bishop method. This method was selected because it is a procedure that is recommended and frequently used in this type of work. While equilibrium conditions are not completely satisfied, this procedure has proven to provide approximately the same results as a more rigorous procedure.[1] A computer program of this method, written in IBM Basic was utilized to analyze this problem. The limitations inherent to the method selected are:

- The results are only as valid as the parameters used for the sludge waste (i.e., unit weight, cohesion, and friction angle).
- unknown water conditions at the site could significantly impact the results.

Four loading configurations were evaluated using this method. Case 1 assumed that Cell A is unsupported and free standing (see figure 3). Case 2 assumed a static surcharge load on the crest of Cell A. Case 3 assumed the same static surcharge load, coupled with a dynamic force impacting Cell A. Case 4 assumed the worst loading conditions; that was a static surcharge load, coupled with a dynamic force, applied to a completely saturated sludge material.

The following assumptions were used in the analysis of these four cases:

- The geosynthetic side liner system was not incorporated into the analysis.
- The dynamic load was modeled in a psuedo-static manner by applying an acceleration term similar to an earthquake term (i.e., E=0.05g).

Table 3 presents the results obtained from this analyses.

Table 3. Slope Stability Results

Case	Type of Loading	Computed F. S.	Type of Failure
1	no external loads	6.3	face
2	static load	3.08	face
3	static & dynamic load	2.68	face
4	static & dynamic load with saturated sludge	1.87	face

NOTE: F.S. $\geq$ 1.5 = acceptable
F.S. $\leq$ 1.5 = not acceptable

Table 3 indicates that this landfilled industrial sludge poses no threat to the stability of Cell A before or after the construction of Cell B. This analysis did reveal, however, that if a failure were to occur, it would most likely occur on the face of the slope.

CONCLUSIONS

The following conclusions may be drawn from this analysis:

- The design methodology presented in this paper proved to be a rational approach for developing a solution to this problem.
- The results of the analysis are only as valid as the laboratory data obtained.
- For this industrial waste sludge, the amount of total expected settlement was tolerable for this multi-cell landfill facility.
- The stability of the north side slope of Cell A appeared to be more than adequate even under extreme loading conditions.
- If a slope failure were to occur, the mode of failure would be a face failure on the north side slope of Cell A.
- A properly designed geosynthetic side slope detail should be able to withstand these calculated settlements. A slope failure, however, may place undesirable mechanical stresses on this system.
- If synthetic systems are employed on the face of a potentially unstable slope, their presence should not be incorporated in the stability analysis.

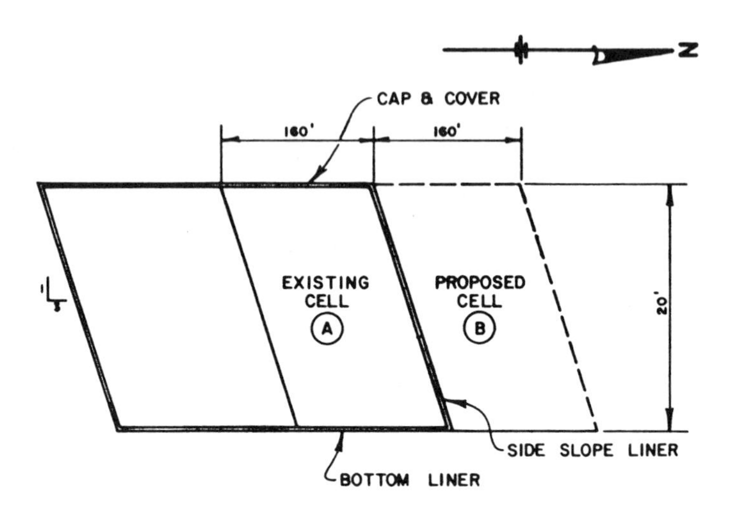

FIGURE (1)

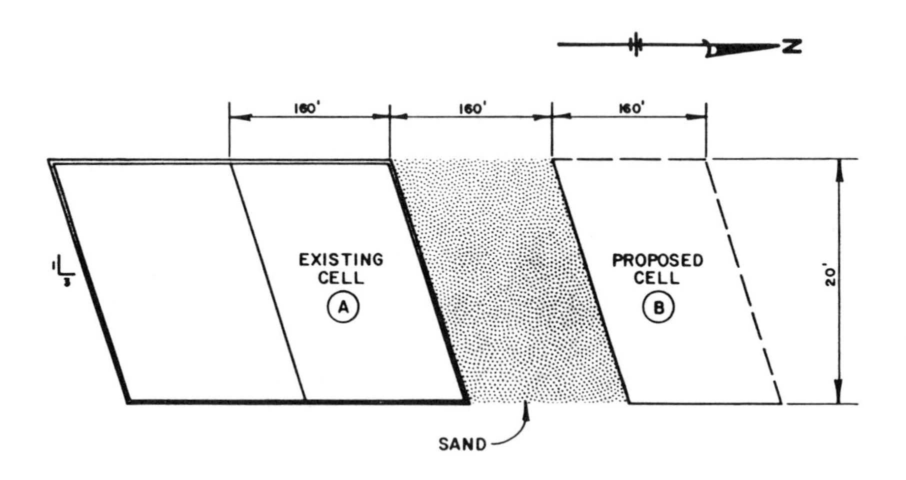

FIGURE (2)

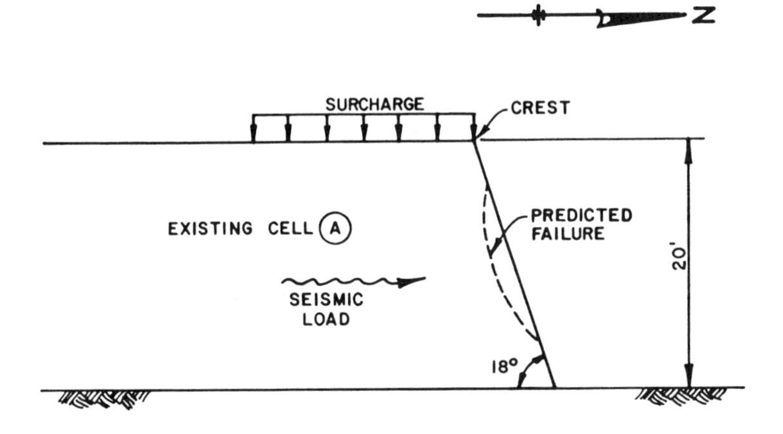

FOUNDATION SOIL

γ = 112 pcf
c = 0
ϕ = 35°
UNSATURATED

FIGURE (3)

References

1. Huang, Y. H. Stability Analysis of Earth Slopes (Van Nostrand, 1983), p. 11.

2. Koerner, R. M. Designing with Geosynthetics (Prentice-Hall, 1986).

3. Das, B. M. Advanced Soil Mechanics (Hemisphere Publishing, 1983).

4. Bowles, J. E. Foundation Analysis and Design (McGraw-Hill, 1982).

LEACHING OF HAZARDOUS AND RADIOACTIVE MATERIALS FROM SOLIDIFIED WASTE FORMS

Kirk K. Nielson and Vern C. Rogers
Rogers & Associates Engineering Corporation,
Salt Lake City, Utah

INTRODUCTION

Many designs for low-level radioactive waste (LLW) disposal include enhanced barriers to contaminant migration. The barriers usually consist of very low permeability materials such as saltstone or concretes. Because of their very low rates of contaminant release, these materials often function as the dominant waste form, and their diffusive transport characteristics dominate the leaching characteristics of the waste in environmental pathway transport codes. This approach provides leach rates that are directly applicable for use in codes using the fractional leach model, such as PRESTO [1], PATHRAE [2], and the NRC's EIS codes in support of 10CFR61 [3]. Diffusion may also be an important mechanism for contaminant transport from the disposal facility to the aquifer in cases where the facility cover permeability is low and the areal extent of the cover is large [4,5].

This paper describes a leaching and transport model that was developed to accommodate diffusion as the dominant transport mechanism. The code treats one-dimensional diffusion through multiple regions, thus dealing with both the enhanced LLW barriers and the low-permeability facility covers that are being considered. Applications of the model are presented in analyses of two typical LLW technology designs.

Geotechnical and Geohydrological Aspects of Waste Management, D. J. A. van Zyl et al., Eds., © 1987 Lewis Publishers, Inc., Chelsea, Michigan—Printed in USA.

LEACH MODEL

The leach model was developed using steady-state diffusion theory, due to the nearly steady-state concentration distributions and fluxes that are established after the first few years in low-permeability systems. These quasi steady-state conditions can be examined from the time-space separated diffusion equations [6,7]. These suggest a relatively simple inverse dependence of contaminant fluxes on the square-root of time for stable chemicals and long-lived nuclides, with more rapid decreases for short-lived species due to decay. The resulting contaminant fluxes, shown generically in Figure 1, give rise to an initial concentration pulse, followed by very slowly-changing concentration distributions that rapidly satisfy steady-state requirements. The low fluxes resulting from the slow diffusional losses also tend to maintain relatively constant source concentrations, further supporting the use of steady-state analysis.

The model is based on molecular diffusion through porous materials. It is designed to apply to water-saturated media such as saltstone and concrete, which derive their water from infiltration and from the surrounding soil. The model was derived as a modification of the RAECOM code [8] for radon gas diffusion through multiple layers of porous materials, and was implemented in a code named RAECORM. The leach rate, as defined by the model, is the annual fraction of contaminant available at the external boundary of

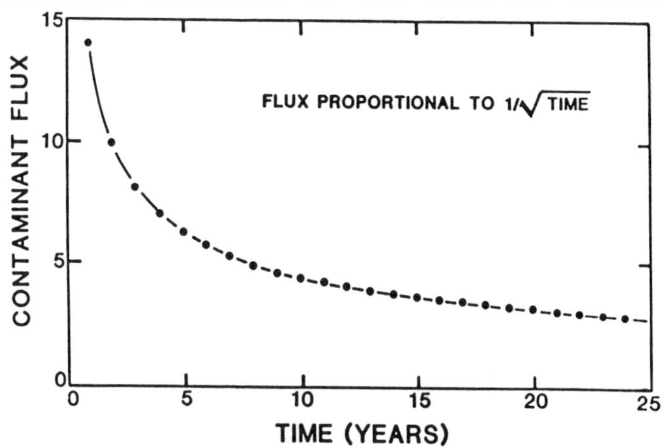

Figure 1. Relative flux of contaminants diffusing from LLW through low-permeability barriers since time of placement.

the LLW containment system. It is defined in terms of the ratio of the total contaminant annual loss to the total quantity of contaminant in the containment system, and hence tends to be constant even after gradual source depletion due to extended diffusional losses.

The following steady-state diffusion equation is solved in RAECORM:

$$D \frac{d^2C}{dx^2} - \lambda C + S = 0, \tag{1}$$

where
- C = contaminant concentration in the total pore space ($Ci\ cm^{-3}$)
- D = diffusion coefficient for the contaminant in the total pore space ($cm^2\ s^{-1}$)
- λ = contaminant decay constant or degradation rate (s^{-1})
- S = steady-state source rate ($Ci\ cm^{-3}\ s^{-1}$)

The flux of contaminant leaving the facility is defined from Fick's law as:

$$J = -Dp \frac{dC}{dx} \tag{2}$$

where
- p = total porosity (dimensionless)
- J = contaminant flux ($Ci\ cm^{-2}\ s^{-1}$)

The leach rate calculated in RAECORM is defined from the concentration profile resulting from Equation 1 and the flux definition in Equation 2 as

$$L = \frac{J \cdot A}{\overline{C} V_s} \tag{3}$$

where
- L = leach rate (yr^{-1})
- A = area for leaching (cm^2)
- $\overline{C}$ = average contaminant concentration in source layer ($Ci\ cm^{-3}$)
- V_s = volume of source layer (cm^3)

The use of Equation 1 is straightforward except in defining the source term, S, to make the steady-state equation apply to the time-dependent source depletion problem. To be applicable to the leaching of contaminants, Equation 1 requires that the annual fraction of source material leached from the facility is small, and that the contaminant spatial distribution is near its equilibrium shape. The first requirement is nearly always met, as it is one of the main reasons for utilizing improved waste forms. By reducing the annual leached fraction to very low levels, the source depletion in one year is negligible and the steady-state equation is applicable. The second requirement is less-directly satisfied at early times, since the relative contaminant concentration distribution changes significantly with time. If initial, uniform concentration distributions are defined throughout the waste volume, the steady-state fluxes and leach rates are significantly less than the large initial spike predicted by the time-dependent solutions. However, comparable fluxes and leach rates are predicted by the time-dependent solutions.

In order to improve the flux and leach rate estimates during the earlier periods, the initial, generally uniform source distribution is altered prior to the steady-state calculations. The alteration involves an extra iteration that is devoted solely to defining a distribution of the source material in the waste volume that is equivalent in total inventory to the specified quantity, but that has a spatial distribution corresponding to the steady-state profile that results after an extended time period. In this way, the contaminant fluxes and leach rates computed for the facility closely correspond to the values predictable from actual contaminant distributions.

In order to compute the steady-state source term distribution, the RAECORM code divides the user-specified source region into at least ten segments and partitions the source distribution among them. It then computes their resulting steady-state concentration profile, and normalizes the source distribution to be identical in shape. Alterations of the source profile due to nuclide decay or chemical degradation are also incorporated in defining the source distribution. The resulting contaminant fluxes are then computed from either the source volume or from transport through any number of containment or soil regions whose properties may each be independently specified. For estimating annual leach fractions, all of the contaminant in the waste volume is assumed to be available for diffusion in each year, and diffusive fluxes are normalized to the spatial average of the contaminant concentration profile, C, as indicated in Equation 3.

APPLICATIONS

The RAECORM leach and diffusive transport model is designed in an general manner, and can be applied to both chemical and radiological contaminants. As examples of the use of the RAECORM model, two applications and their results are presented here. The first example is that of radioactive LLW that has been processed and cast into a saltstone monolith with the nominal dimensions of a 55-gallon disposal drum. The second example is identical except that it also includes a concrete enclosure around the waste.

The characteristic radius of the saltstone monolith for one-dimensional representation was estimated to be 117 cm, with diffusion coefficients, porosities, and moistures as indicated in Table 1. The resulting steady-state concentration profiles computed by RAECORM are represented by the lower curve in Figure 2. The boundary of the saltstone for this case was a soil material, whose diffusion and permeability characteristics were sufficiently non-restrictive to permit escaping waste to be rapidly removed from the saltstone. Hence, its boundary concentration remained near zero. The resulting contaminant flux and leach fraction determined for the saltstone are presented in Table 2. The indicated annual leach fraction of 8.6 x 10^{-5} y^{-1} is well within the realm of the steady-state and non-depleting source assumptions, and is suitable for the thousands-of-year time frames contemplated for performance assessments on LLW disposal technologies.

Table 1. Defining Parameters for Leaching and Transport Calculations From Saltstone and Concrete Containment Materials

	Saltstone	Concrete Containment
Diffusion Coefficient	5.0E-8 cm^2 s^{-1}	5.0E-7 cm^2 s^{-1}
Total Porosity	0.3	0.3
Moisture (dry wt)	15.7%	15.7%
Characteristic Thickness	117 cm	61 cm

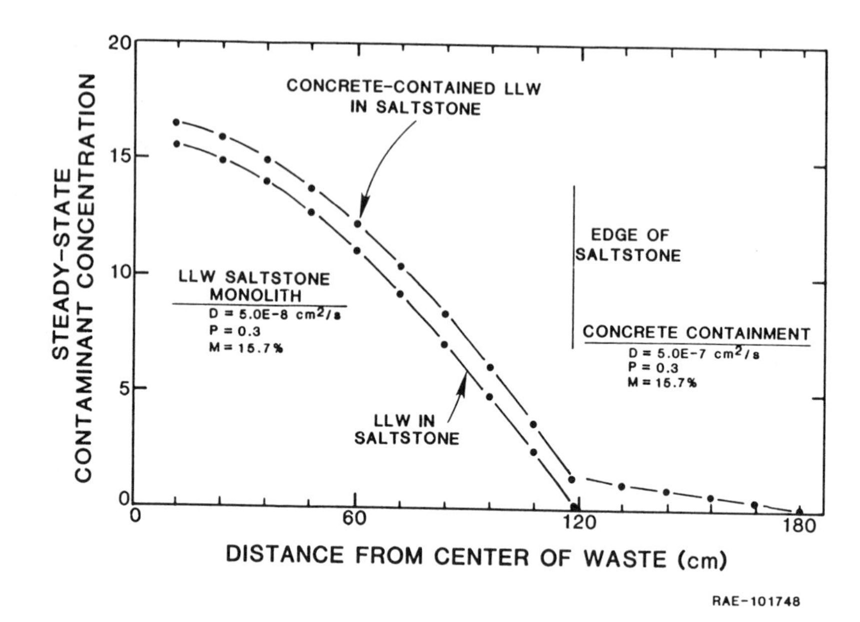

Figure 2. Steady-state contaminant concentration profiles in LLW saltstone monolith with and without concrete containment.

Table 2. Results of Leaching and Transport Calculations for Saltstone and Concrete-Contained Saltstone LLW Monoliths.

	Saltstone	Concrete-Contained Saltstone
Contaminant Flux[a]	3.2E-4 cm^2 s^{-1}	3.2E-4 cm^2 s^{-1}
Annual Leach Fraction	8.6E-5 y^{-1}	7.7E-5 y^{-1}

The second example involves a two-foot thick concrete containment, which serves to contain some of the contaminants before release into the faster-moving groundwater/aquifer system. As illustrated by the upper curve in Figure 2, this results in a very similar contaminant concentration distribution in the source region, that is increased over that in the first example by an increment defined by the diffusion coefficient and thickness of the concrete. This behavior results from the ten-fold higher diffusion coefficients of the concrete (Table 1), which still allow the diffusion in the saltstone to dominate. For LLW in unimproved waste forms with high diffusion coefficients, the two-foot concrete containment causes a more marked increase in the concentration profiles in the waste region. Annual leach fractions for the saltstone/concrete case are also shown in Table 2, and as expected, show slightly lower values than those for the bare saltstone. Within the calculated precision, fluxes are virtually identical for the two cases.

The RAECOM model also accommodates transport calculations for species that are defined to have retardation coefficients, or to have a reduced diffusion rate in any or all of the regions of the system. The reduced diffusion rates may result from ion exchange, surface adsorption, or other phenomena. For these cases, the diffusive transport through all regions exhibiting the retardation behavior can be exactly represented by simply dividing the characteristic diffusion coefficient for the region by the corresponding retardation coefficient.

Applications of the RAECORM model to LLW facilities and waste forms of various sizes has been used to evaluate monolith dimensions, packaging, and containment options. Characteristic dimensions are determined as the average of the half-thicknesses weighted by the perpendicular surface areas for each orthogonal axis. These analyses indicate that short characteristic path lengths lead to dramatic increases in effective leachability, even for improved waste forms such as saltstone. These are summarized by the leach fractions in Table 3. For a simple one-region containment system, the leach rate is inversely proportional to the square of characteristic thickness. The estimates in Table 3 also suggest that very good contaminant isolation can be achieved if sufficiently large LLW monoliths or low-permeability containments can be devised that will withstand mechanical forces that would reduce their effective dimensions.

Table 3. Comparative RAECORM Leaching Calculations for LLW Facilities of Various Characteristic Sizes.

Characteristic[a] Dimension (cm)	Annual Leach[b] Fraction (y^{-1})
10	.012
30	.0013
100	.00012
300	.000013
1000	.0000012

REFERENCES

1. "PRESTO-EPA-POP: A Low-Level Radioactive Waste Environmental Transport and Risk Assessment Code -- Methodology Manual," U.S. Environmental Protection Agency Report EPA 520/1-85-001 (1985).

2. Merrell, G.B., V.C. Rogers, K.K. Nielson, and M.W. Grant, "The PATHRAE Performance Assessment Code for the Land Disposal of Radioactive Wastes," Rogers and Associates Engineering Corporation Report RAE-8469/3 to U.S. Environmental Protection Agency (1985).

3. Widmayer, D.A., "User's Guide for 10CFR61 Impact Analysis Codes," U.S. Nuclear Regulatory Commission Report NUREG-0959 (1983).

4. Dougherty, D.R., M. Fuhrmann, and P. Colombo, "Accelerated Leach Test Program Annual Report," Brookhaven National Laboratory Report BNL-51955 (1985).

5. MacKenzie, D.R., J.F. Smalley, C.R. Kempf, and R.E. Barletta, "Evaluation of the Radioactive Inventory in, and Estimation of Isotopic Release From, the Waste in Eight Trenches at the Sheffield Low-Level Waste Burial Site," U.S. Nuclear Regulatory Commission Report NUREG/CR-3865 (1985).

6. Bell, M.J., "An Analysis of the Diffusion of Radioactivity From Encapsulated Wastes," Oak Ridge National Laboratory, Chemical Technology Division, Contract W-7405-eng-26 Report (1971).

7. Petschel, M., and D. Zappe, "On the Applicability of Analytical Solutions of the Transport Equations for the Migration of Substances in Groundwater-Bearing Horizons," Nuclear Technology 65, pp. 125-130 (1984).

8. Rogers, V.C. and K.K. Nielson, "Radon Attenuation Handbook for Uranium Mill Tailings Cover Design," U.S. Nuclear Regulatory Commission, NUREG/CR-3533 (1984).

ASSESSMENT OF THE EFFECTS OF POTENTIAL GROUND SUBSIDENCE ON A RECLAIMED TAILINGS PILE

Jack A. Caldwell and Ned Larson, Jacobs Engineering Group Inc., Albuquerque, New Mexico

Ron Rager, Sergent, Hauskins, & Beckwith, Geotechnical Engineers, Albuquerque, New Mexico

INTRODUCTION

The UMTRA Project involves remedial work at 24 abandoned uranium mill tailings piles. Remedial work is to be effective for a design life of 1000 years, where reasonably achievable, and at any rate for 200 years.

At the Ambrosia Lake pile in New Mexico, there are mine workings at about 174 to 180 meters below the southwest part of the pile. Potential subsidence, if it occurs, could impact the integrity of the remedial work to be undertaken at the pile.

This paper describes the evaluations and analyses done to assess the potential for and nature of subsidence of strata beneath the pile. Further, the paper describes the work done to assess the impacts, if any, of potential subsidence on the pile and in particular on the radon barrier - a cover of compacted silty clay placed over the pile.

The analyses show that even if the maximum theoretically predicted subsidence were to occur, there will be no significant effect on the integrity of remedial works at the pile.

SITE DESCRIPTION AND GEOLOGIC SETTING

The Ambrosia Lake site is in northwest New Mexico, within the Navajo section of the Colorado Plateau

Geotechnical and Geohydrological Aspects of Waste Management, D. J. A. van Zyl et al., Eds., © 1987 Lewis Publishers, Inc., Chelsea, Michigan – Printed in USA.

physiographic province (Figure 1). Terrain in this section of the plateau is generally flat-topped, gently sloping cuestas, broad steep scarped mesas, low-gradient pediment and fan surfaces, deeply incised canyons and arroyos, and strike valleys. Basalt flows and cinder cone fields cover large areas west and south of the Ambrosia Lake site.

Cenozoic tectonic and erosional processes have exposed rocks of Precambrian through late Quaternary age in the region. With the exception of Quaternary basalt flows, Precambrian igneous and Paleozoic sedimentary units outcrop south of the tailings site in the vicinity of the Zuni Uplift. Progressively younger strata are exposed in the central and northern portions of the area, terminating with deltaic deposits of the upper Cretaceous Fruitland Formation. Extrusive volcanic rocks primarily of Miocene and Pliocene age constitute the Mt. Taylor and Mesa Chivato physiographic features east and northeast of the existing tailings pile.

The Ambrosia Lake site lies in a northwest-trending strike valley cut into the upper Cretaceous Mancos Shale. Approximately 1000 meters of Permian to Cretaceous clastic sedimentary strata underlie the site. The sedimentary section dips northeast at approximately two degrees, forming a regional homocline of the southern San Juan Basin referred to as the Chaco Slope.

Tectonic faults are abundant in the study region and reflect multiple episodes of deformation. Most of the structures are Laramide, with north and northeast trends. Displacement is commonly down to the east and on the order of a few tens of meters. In addition to the regional uplifts and monoclinal elements forming the southern margin of the San Juan Basin, numerous small-scale domes, anticlines, and synclines locally deform the otherwise uniform regional bedding. Vertical displacement of the Laramide age local structures is generally less than 150 meters.

Unconsolidated alluvial and eolian deposits of late Quaternary age mantle extensive low-lying portions of the study region and site. Within the Ambrosia Lake valley, thicknesses of alluvium exceed 30 meters. Valley sideslope alluvial sediments in the immediate site vicinity range in depth from one to 15 meters.

THE PROPOSED REMEDIAL ACTION AND DESIGN FEATURES

The main feature of the design concept is the consolidation and stabilization in place of the Ambrosia Lake tailings, contaminated subsoils, and windblown contaminants. Mill buildings and foundation materials will also be demolished and buried.

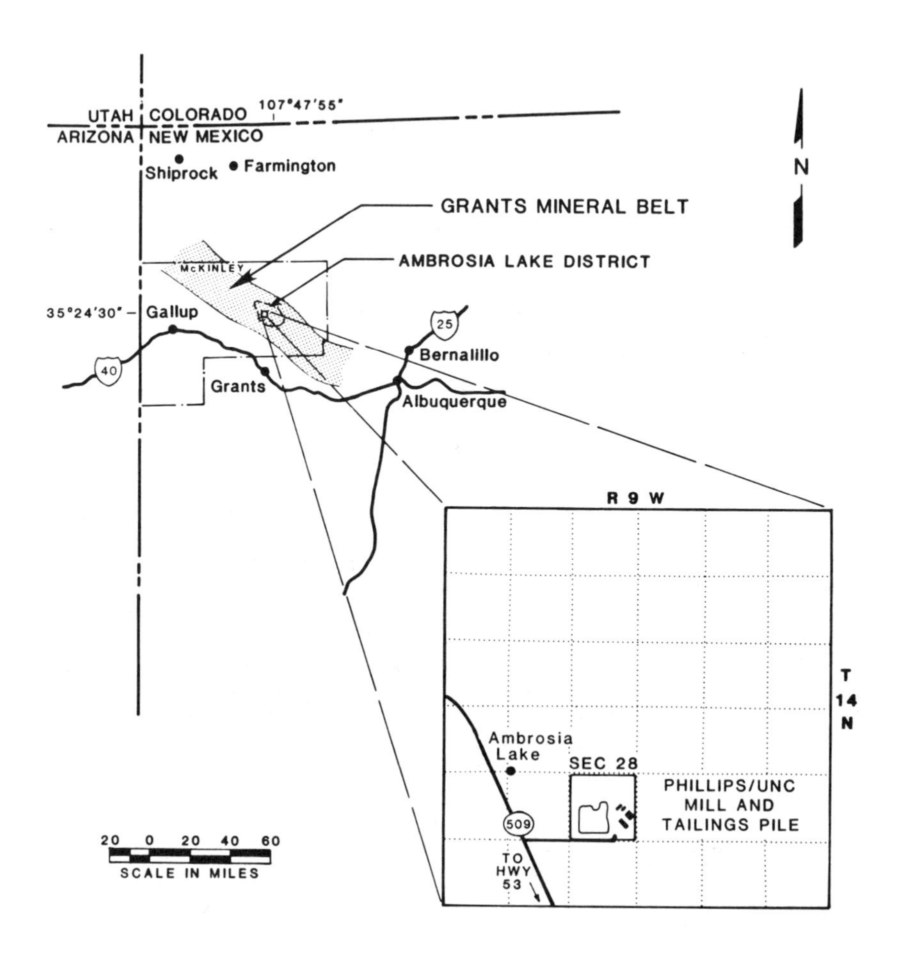

Figure 1. Vicinity map of Ambrosia Lake UMTRA site.

Figure 2 shows the cross-section of the pile and Figure 3 shows a layout of the pile after remedial work. Figure 3 also shows the zone of mine workings below the southwest corner of the pile which could lead to subsidence.

The radon barrier is compacted weathered Mancos Shale from the borrow site approximately 1.5 kilometers north of the pile. The radon barrier, which is at least one meter thick, is designed to reduce radon flux to 20 picocuries per square meter per second or less.

Pile erosion protection consists of rock and filter layers designed to withstand the Probable Maximum Precipitation (PMP) on the pile and the Probable Maximum Flood (PMF) flows around the perimeter of the pile.

The northern portion of the pile is to be relocated because it is more economical to move the tailings and incorporate them into the main portion of the pile, than it is to cover them with a radon barrier and erosion control layers. The tailings moved from the northern portion of the pile will be used to fill the existing pond on the pile surface and to shape the top of the pile to form slopes from which water will drain.

The southern portion of the pile, including the southwestern corner which is underlain by mine workings, will not be moved as it is more economical to stabilize this portion in place and cover it with a radon barrier and erosion control layer rather than to relocate it.

Tailings instability is another reason for not relocating the tailings in the southwestern part of the pile. The tailings in that part of the pile are soft and nearly saturated. If the surrounding sand tailings dike were to be removed, problems of instability would be encountered: the soft slimes tailings in the inner part of the pile could flow from a breach in the perimeter dikes.

REGIONAL SUBSIDENCE

Two principal mining methods were used in the extraction of the Ambrosia Lake ore. A technique referred to as "scram" stope mining was used to exploit the thickest ore bodies during the early years of activity in the Ambrosia Lake valley. Changes in mining industry regulations and the decreasing grade of ore prompted mining firms to develop ore bodies during the 1970s using the multiple level room and pillar method. Two levels of room and pillar type workings underlie the southwest corner of the tailings pile at a depth of approximately 174 to 180 meters.

Numerous occurrences of mine working-induced subsidence have been recorded in the Ambrosia Lake

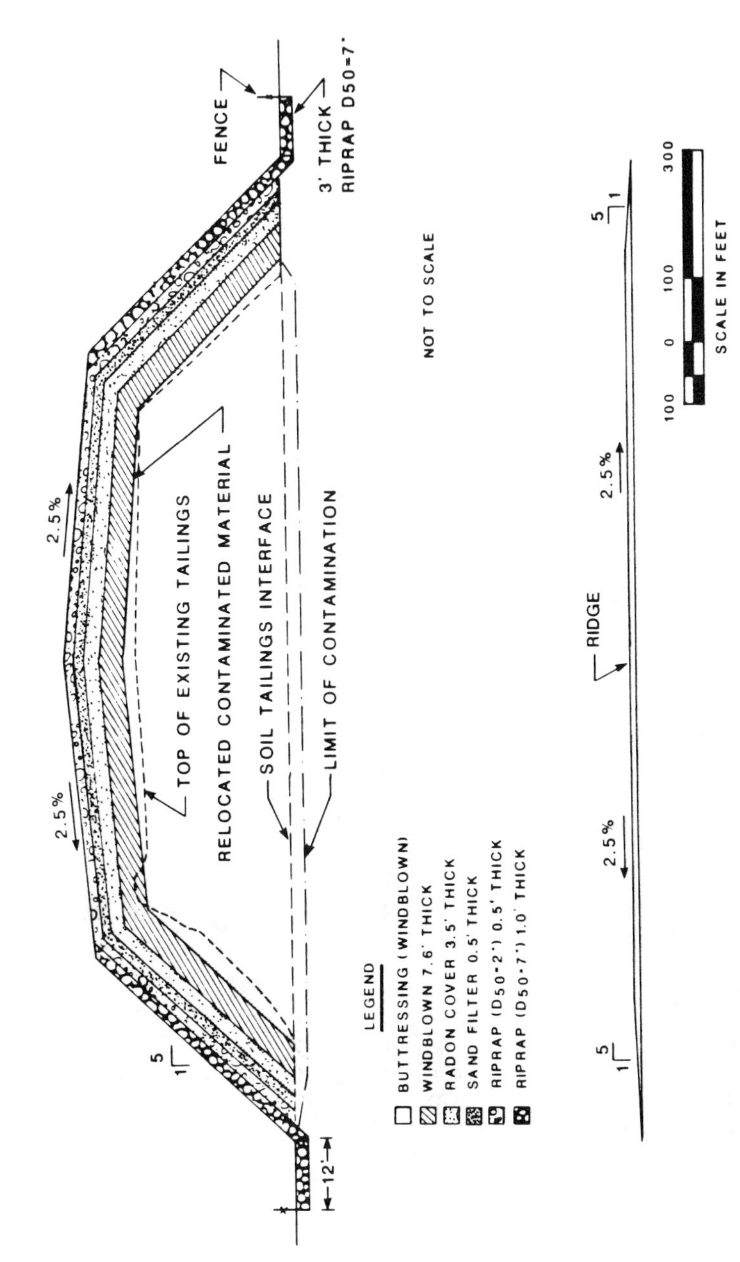

Figure 2. Typical embankment cross-section, Ambrosia Lake site.

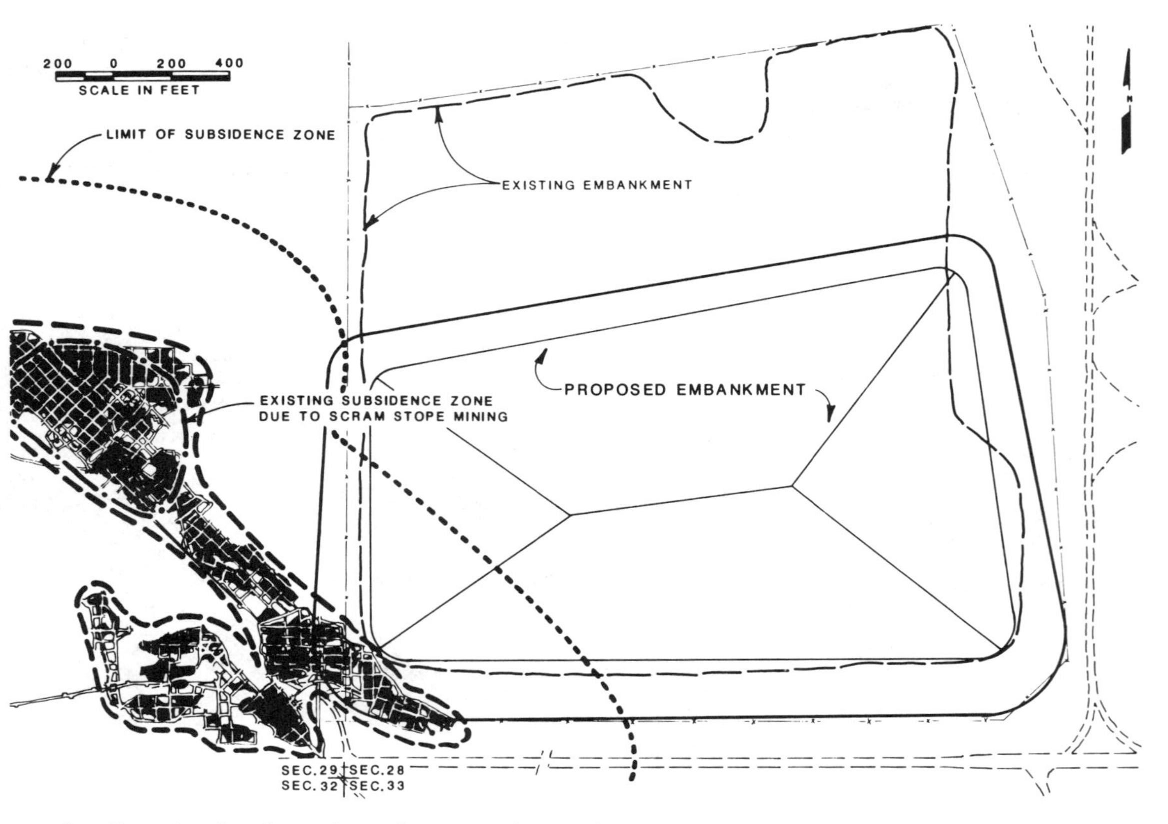

Figure 3. Map showing location of proposed embankment and subsidence zone

mining district. Surface manifestation of the collapse of underground workings is evident approximately 250 meters west of the existing tailings pile (Figure 3).

Close examination of mine workings maps provided by firms operating in the Ambrosia Lake area indicates that surface subsidence is nearly always associated with "scram" stoping development of the underlying ore body. As the mining technique involved the use of the entire rock column overlying the ore bearing strata as a natural gravity crusher of the ore, failure of the ground surface was concurrent with mining activity. The depressions are circular or elliptical and range from a few to several meters deep. The zone of subsidence on the western edge of the existing tailings pile is a depression approximately 400 meters in diameter, and one meter deep at the lowest point. The subsidence zone is delineated by discontinuous subsidence cracks, piping holes, and enhanced vegetation growth along its perimeter. There are two concentric subsidence scarps of two- and one-meter widths along the north and east edges of the subsidence zone. Erosion and piping have developed troughs in the scarps exceeding a depth of one meter, although vertical offset of individual scarp features is not greater than 150 millimeters. Similar zones of subsidence are seen in airphotos of the Ambrosia Lake valley.

No recognized surface subsidence has occurred over any areas of multiple room and pillar development in the Ambrosia Lake area. Abel and Lee (1) in a study of the lithologic controls and subsidence observed that, in Pennsylvania, the negligible magnitudes of surface subsidence associated with the failure of room and pillar type workings were frequently undetectable. This appears to be true also for areas of room and pillar mining in the Ambrosia Lake ore district.

Sandstones, siltstones, and shale of Jurassic and Cretaceous age constitute the 200-meter-thick rock column overlying the sub-pile mine voids. No known tectonic structures disturb the uniformly dipping units. The ore-bearing strata are poorly to very poorly indurated, and presented a constant caving hazard in the workings during the years that mines were operating. Ground support of mine passages was typically placed on centers of two meters or less. Ceiling failure generally occurred in the form of localized arcuate slabs. It is possible, therefore, that collapse of the stopes has already occurred. Peele and Church (2) describe early studies in which the volumetric expansion that desegregated rock slabs underwent during the failure of successively shallower strata were characterized. The phenomenon, termed "swell factor," varies primarily with the lithologic character of the rock column and depth of workings being evaluated. Application of the values quoted by Peele and Church (2) suggests that complete infilling of the failure void would occur at a depth of 150 meters below ground surface. The analysis assumes a vertical dimension of mine opening roughly three times that beneath the tailings and is therefore a conservative approach.

VERTICAL MOVEMENTS AND EXTENT OF SUBSIDENCE

The methods used to analyze the subsidence are described by Abel & Lee (1) and the National Coal Board (NCB) (3). The mining under the pile consists of rooms and pillars. The geology overlying the underground openings is described above and shown in Figure 4. The characteristics of each opening are given in Table 1.

Table 1. Mine Openings Characteristics.

	Upper opening	Lower opening
Depth	174 m	180 m
Panel width	30 m	36 m
Panel length	245 m	245 m
Mining height	4.5 m	4.5 m
Pillar width	3.0 m	3.0 m
Extraction ratio	60%	70%

The maximum subsidence was calculated using the equation presented by Abel & Lee (1) as follows:

$$\text{Subsidence } (\%) = 8.469 + 11.95\ln(L_{max}\ (\frac{H}{W}))$$

$$L_{max} = \frac{KD}{1-R}$$

where:

D = depth (m).
R = extraction ratio.
H = mining height (m).
W = pillar width (m).
K = A constant-the unit mass of rock-0.0226 MPa/ft.

The maximum percent subsidence was calculated to be 40.6 percent for the upper level and 44.5 percent for the lower level. This results in a critical subsidence for the upper and lower levels of 1.86 and 2.03 meters, respectively. Since the panels are fairly narrow with respect to the depth, the subsidence is sub-critical and a reduction in the critical subsidence may be made. NCB (3) quotes reduction factors to account for this;

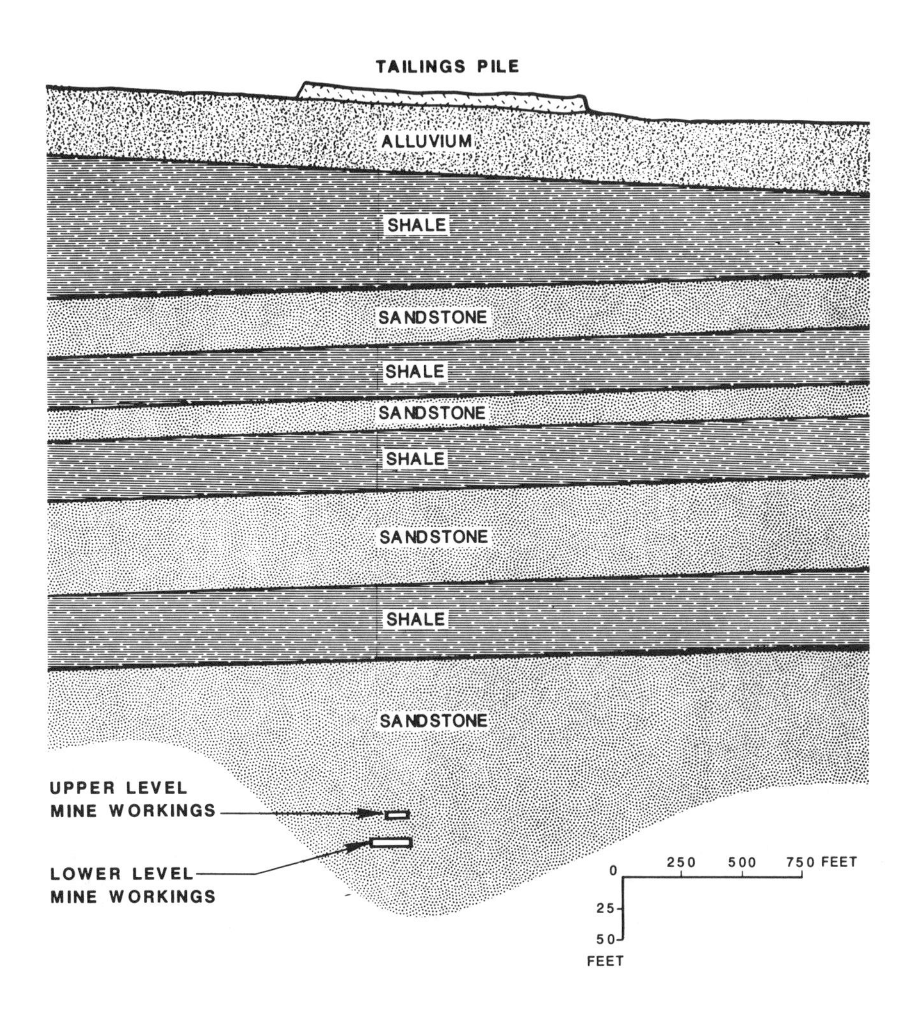

Figure 4. Geology above and location of underground mine workings.

values of 0.04 and 0.05 for the upper and lower levels, respectively, were used in the analysis. This produces an estimated settlement of 73 and 101 millimeters for for the upper and lower levels with a combined total subsidence at the surface of 174 millimeters (4).

In order to predict the profile of the subsidence, the angle of draw is determined using empirical relationships presented by Abel & Lee (1). The materials above the workings consist of about 52 percent shale and 48 percent sandstone. For this lithology, the angle of draw is 20°.

The subsidence profile was calculated and plotted as described by the NCB and shown in Figure 5.

RADON BARRIER CRACKING

Subsidence itself is not necessarily harmful to the radon barrier but the amount of strain that the radon barrier experiences is critical because too much tensile strain could cause cracking. Compressive strain in the radon barrier would not necessarily be harmful because cracking would not occur and the radon barrier is very plastic; therefore, a shortening would be expected to occur but no buckling is likely.

Since the strain is the first derivative of the subsidence profile, the strain can also be plotted. The guidelines shown by the NCB were also used to plot the strain profile. Figure 6 shows the strain profile for the combined workings. The maximum tensile strain was calculated to be 0.05 percent.

This strain can be compared to values published for structural damage. Cracking of plaster occurs for strains of 0.1 percent, and cracking of reinforced concrete occurs at about 0.3 percent.

Leonards and Narain (5) report an extensive series of tests to measure the strains that actually cause cracking in soils subjected to tensile stresses. Some of the soils tested are similar to those to be used for the radon barrier at Ambrosia Lake. Their testing indicates that cracking does not occur in compacted earthen structures until strains of up to 0.3 percent are reached. This indicates that even the maximum strains that could potentially occur at the site are lower than those at which cracking is likely to occur.

Localized cracking or failure is not likely to occur at this site since no major faults or changes in lithology occur within the potential subsidence zone.

The discussion above is based on the approach that the subsidence will be expressed at the surface as a relatively uniform downward movement and smooth deformation profile. The basis for this is the presence of a layer of Mancos Shale above the workings and beneath the pile. The Mancos Shale is a relatively uniform material with no significant structure.

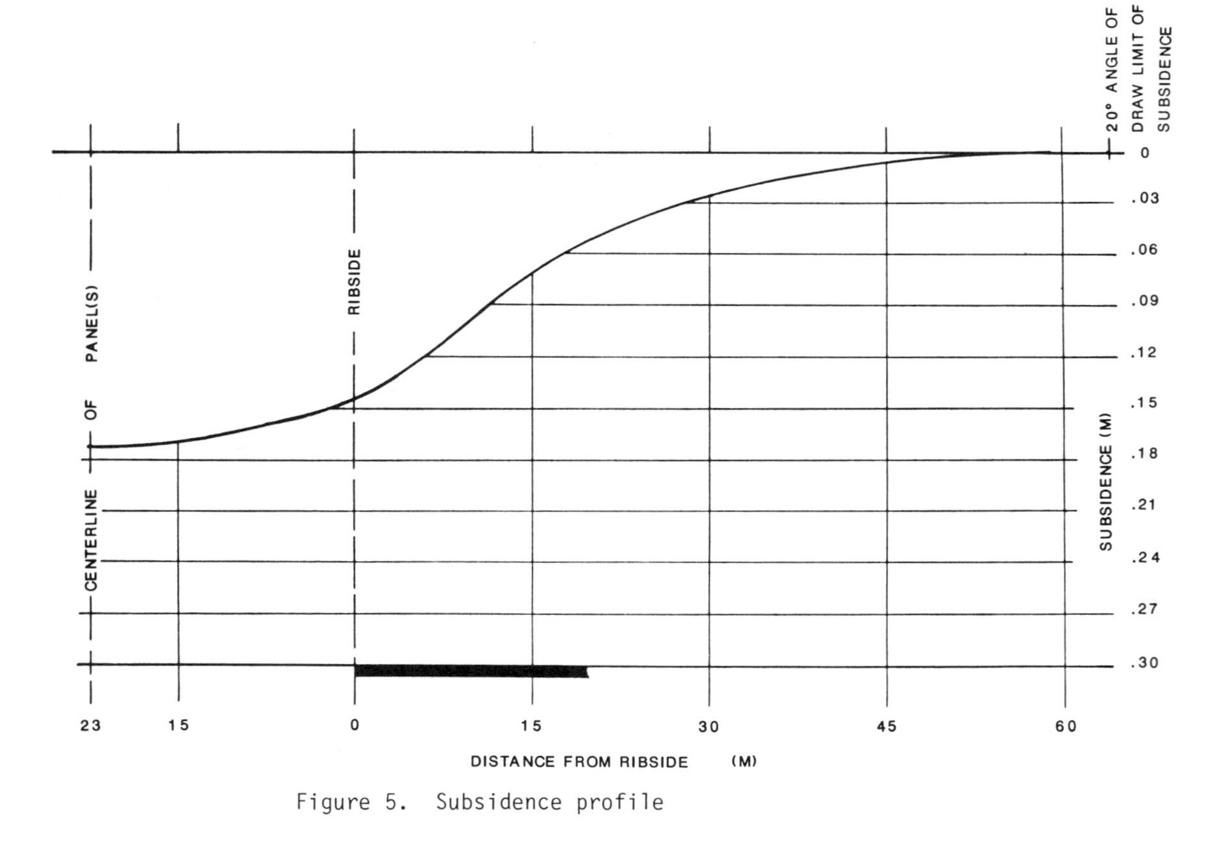

Figure 5. Subsidence profile

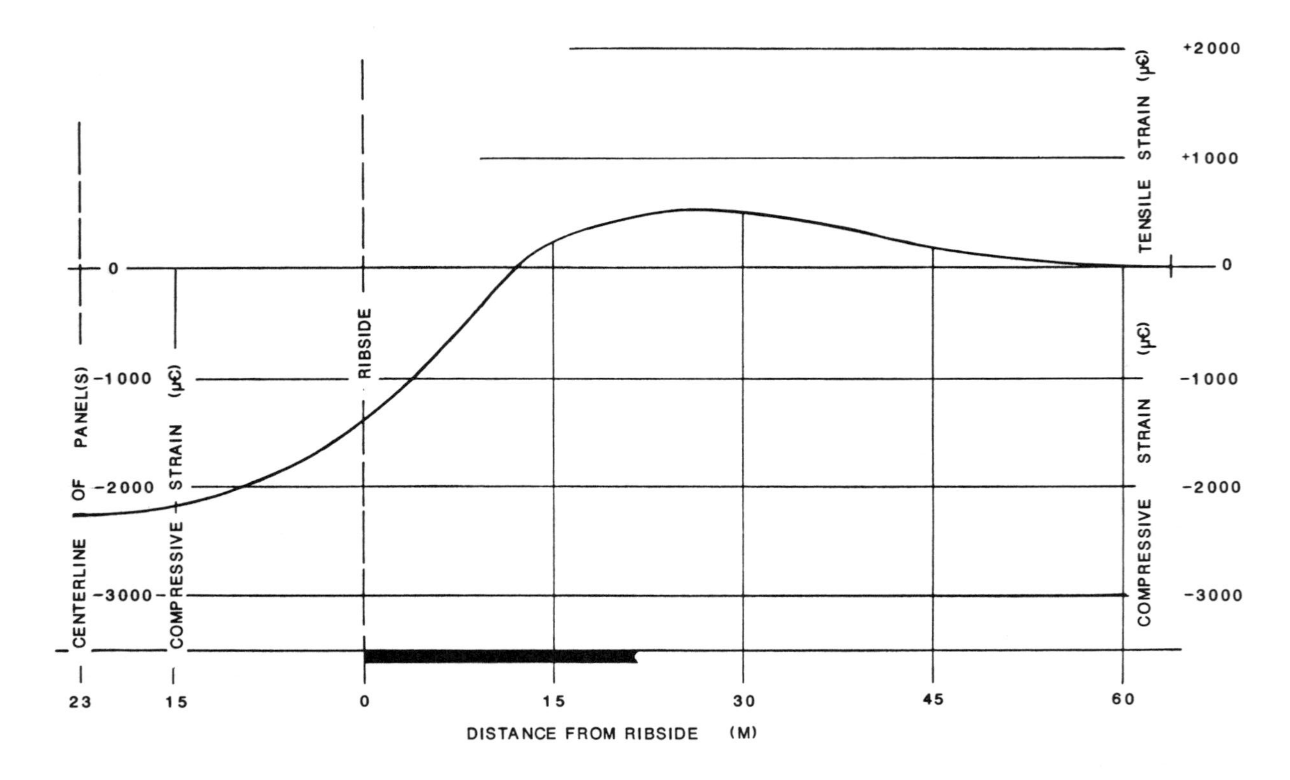

Figure 6. Strain profile

There is, however, a remote possibility that subsidence will occur as a distinct surface step at a specific location. The effect of this could potentially be a localized shear deformation of the cover. This would be manifest through the rock armor at the surface. A single or several shear and extensional cracks will have little effect on the intended performance of the radon cover. In addition, the rock armor that protects the pile from surface water runoff is not significantly affected since the subsidence scarp would be small and perpendicular to any runoff flow path.

CONCLUSION

Subsidence does not always occur over all mine openings. Localized collapse of roof strata can effectively seal a mine opening due to the increased volume of broken rock compared to in-situ rock. This change in volume is called the swell factor, and thus may prevent or reduce subsidence.

For the Ambrosia Lake inactive uranium mill tailings piles, the maximum vertical displacement, horizontal movement, and strain of the ground surface were calculated. The strains induced in the radon barrier, a silty clay cover placed over the pile, were calculated. The maximum predicted strain in the cover was compared to the established limits for compacted earthen structures, and found to be below the limit at which cracking will occur.

Accordingly the conclusion is that subsidence, even if it does occur, is unlikely to affect the integrity of the remedial works at the pile. Hence, it is not necessary to relocate the tailings at the southwest corner of the pile. This is a cost-effective solution and one that is also acceptable for practical reasons.

ACKNOWLEDGMENT

This study was completed as part of the Department of Energy's Uranium Mill Tailings Remedial Action Project, headquarted in Albuquerque, New Mexico, and was supported under DOE Contract No. DE-AC04-82AL14086 to Jacobs Engineering Group Inc.. The Jacobs-Weston Team consists of Jacobs Engineering Group Inc., Roy F. Weston, Inc., and Sergent, Hauskins & Beckwith Geotechnical Engineers, Inc.

REFERENCES

1. Abel, J. F., Jr., and F. T. Lee. "Lithologic Controls on Subsidence," in Proceedings of the American Institute of Mining Engineers (AIME), Vol. 274 (Littleton, Colorado, 1977).

2. Peele, R., and J. A. Church. Mining Engineering Handbook (New York: John Wiley & Sons, Inc., 1975).

3. NCB (National Coal Board). Subsidence Engineers' Handbook (London: National Coal Board, 1975).

4. Abel, J. F. Unpublished report for Jacobs Engineering Group Inc., Albuquerque, New Mexico, (1986).

5. Leonards, G. A., and J. Narain. "Flexibility of Clay and Cracking of Earth Dams," Journal of the Soil Mechanics and Foundation Division (American Society of Civil Engineers, 1975).

METHODOLOGY FOR OVERSIZING MARGINAL QUALITY RIPRAP FOR EROSION CONTROL AT URANIUM MILL TAILINGS SITES

William P. Staub,
Oak Ridge National Laboratory, Oak Ridge, Tennessee

Steven R. Abt
Colorado State University, Fort Collins, Colorado

ABSTRACT

Properly selected and oversized local sources of riprap may provide superior erosion protection compared with revegetation at a number of uranium mill tailings sites in arid regions of the United States. Whereas highly durable rock is appropriate for protecting diversion channels to the height of the 5-year flood, marginal quality rock may be adequate for protecting infrequently flooded side slopes of diversion channels, tailings embankments and caps. Marginal quality rock may require oversizing to guarantee that design size specifications are met at the end of the performance period (200 to 1000 years).

This paper discusses a methodology for oversizing marginal quality rock. Results of cyclic freezing and thawing tests are used to determine oversizing requirements as functions of the performance period and environment.

Test results show that marginal quality rock may be used in frequently saturated areas but in some cases oversizing will be substantial and in other cases marginal quality rock may be disqualified. Oversizing of marginal quality rock appears to be a practical reality in occasionally saturated areas (between the 5-year and 100-year floods). Furthermore, oversizing will not generally be required on slopes above the 100-year flood.

INTRODUCTION

Protection of the public health and environment from potential hazards of radioactive waste and uranium mill tailings has

Geotechnical and Geohydrological Aspects of Waste Management, D. J. A. van Zyl et al., Eds.,

stimulated the assessment of waste stabilization procedures and methods. Under current stabilization procedures, mill tailings embankments and caps are allowed to dry out sufficiently to permit the use of heavy construction equipment in their reclamation [1]. Diversion channels protected with riprap are constructed around the tailings to reduce natural drainage across them. The drained tailings are graded and capped with an earthen cover. The engineering design of these covers should provide site stability with little or no maintenance and should insure long-term integrity for 200 to 1000 years. [2]

Impoundment covers are generally comprised of native soils. The emplaced earthen cover is more vulnerable to sheet, rill and gully erosion than undisturbed soil with similar site conditions. Therefore, it is imperative that erosion prevention measures be considered in the design process to minimize cover degradation.

The placement of a rock armor or revetment over the cover is an alternative to the use of vegetation for protecting the waste impoundment. The rock is durable and erosion resistant. Once the armor layer is placed, long-term maintenance programs are usually not required. Therefore, rock protection can be considered a viable alternative for long-term protection of waste storage sites where usable riprap sources are present.

Riprap selection may be based on methodologies developed by Foley et al. [3] and Nelson et al. [4]. Riprap testing consists of a sequence of physical and petrographic tests which can be used to rank rock quality as good, fair, marginal, or unsuitable according to a numerical ranking system.

In some cases, fair and marginal quality rock may be oversized to provide assurance that it will meet design criteria throughout the performance period. This paper discusses a methodology for oversizing marginal quality riprap for use in stabilizing the cover over dry mill tailings. This methodology would not be suitable for oversizing riprap for operating mill tailings ponds or conventional reservoirs where standing bodies of water may effect its long-term durability.

OVERSIZING METHODOLOGY

Riprap oversizing is based on the assumption that the principal failure mechanism is cyclic freezing and thawing. The amount of oversizing is a function of the number of freezing and thawing cycles expected during the performance period as determined from meteorological data and the percent weight loss per cycle as determined by a standardized freezing and thawing test. Assuming a linear rate of weight loss as a function of the number of freezing and thawing cycles, oversizing can be calculated from Equation 1, assuming an even distribution of weight loss around the surface of a sphere:

$$\text{Eqn (1)} \quad r_o^3 = \frac{100}{(100 - wN)} \cdot r_d^3 ,$$

where r_o and r_d are oversizing and design radii, respectively; w is the percentage weight loss per freezing and thawing cycle and N is the estimated number of cycles in the performance period.

Riprap durability requirements are a function of the riprap location relative to surface water flow channels. Riprap placed in frequently flooded locations is most vulnerable and its performance is most critical to the overall embankment stability. If all elements of the tailings containment structure are properly designed and constructed, surfaces will be well-drained so that only materials in the diversion channel will be periodically saturated for prolonged intervals of cyclic freezing and thawing on an annual basis.

APPLICATION OF THE METHODOLOGY

Several riprap sources were used to investigate riprap deterioration as a result of freeze-thaw cycling. Quarried, unweathered granitic rocks from scattered regions in Wyoming were collected and tested over the last 40 years by the U.S. Bureau of Reclamation [5]. Additional samples were collected and tested for this study, including granitic boulder conglomerates from a mine in the East Gas Hills (East Canyon conglomerate), from an outcrop in the West Gas Hills (Dry Coyote conglomerate), Wyoming, and sandstones (Wasatch Formation) from a mine and outcrops in the Powder River Basin, Wyoming. All the granitic rocks rang when struck with a hammer and were extremely difficult or impossible to break. They ranged in size from cobbles to boulders. Most of the sandstones also rang when struck with a hammer but they were somewhat less difficult to break. The sandstones were boulder sized concretions that were well cemented with calcite and, in some cases, ferruginous cement. Blasting had been required to dislodge samples that were taken from a mine. Petrographic descriptions [6] indicate the amount of smecktite clay minerals present in the sandstones was less than 2%. Based on the above information, it is believed that all these samples would perform satisfactorily in a slaking and abrasion test [3].

For purposes of demonstrating the application of the riprap oversizing methodology presented here, it is assumed that a hypothetical region has 40 complete freeze-thaw cycles per year. Thus,

1) in the area which is below the annual flood elevation and fully saturated conditions exist, riprap would be subjected to 40 annual complete freeze-thaw cycles, and
2) riprap above the 5-year flood and 100-year flood have one-fifth and one-hundredth as many complete freeze-thaw cycles over the long-term as environments that are annually flooded.

Table 1 lists the number of expected cycles (N), below the annual flood and adjacent to the 5-year and 100-year floods, for 200-year and 1000-year performance periods, respectively, using the foregoing assumptions.

The standardized 250 cycles freeze-thaw test was used to determine the percent weight loss per test and per cycle for each riprap source (Table 2). Sodium sulfate soundness and Los Angeles

Table 1. Estimated number of freeze-thaw cycles (N) for various flood levels and performance periods (assuming the annual number of cycles = 40 below the annual flood in a hypothetical region).

Performance Period	Flood Recurrence Interval		
	Annual	5-year	100-year
200 years	8,000	1,600	80
1000 years	40,000	8,000	400

abrasion tests were also included for comparison. From a comparison of the test results in Table 2 with the USBR standards in Table 3 for judging riprap durability, quarried, unweathered granitic rocks performed well in all three tests. Granitic boulder conglomerates were intermediate to borderline good in quality based on sodium sulfate soundness and Los Angeles abrasion tests and good quality based on freezing and thawing tests. On the otherhand, concretionary sandstones were extremely poor to poor quality in sodium sulfate soundness and Los Angeles abrasion tests, respectively, while they were intermediate in quality based on freezing and thawing tests. Hence, a full range of riprap quality is provided by the source samples indicated in Table 2.

Equation 1 and data from Tables 1 and 2 were used to calculate oversizing ratios (r_o/r_d) for various riprap samples. Table 4 shows oversizing ratios as a function of performance period and flood recurrence interval. Based on Equation 1, all of the sandstone samples would have completely disintegrated over a 1000-year performance period when subjected to annual flooding. Calculations suggest that two of these sandstones might last 400 years with oversizing ratios ranging between 1.22 and 2.13. According to this methodology only moderate oversizing would be required for concretionary sandstones placed above the 5-year flood and no oversizing would be required above the 100-year flood.

Surprisingly, Eocene boulder conglomerates performed as well as quarried unweathered granitic rock in the freezing and thawing test. These rocks require only nominal oversizing when subjected to annual flooding, very little for flood intervals up to 5 years, and none at all above the 5-year flood.

CONCLUSIONS

As illustrated by the above examples, freezing and thawing tests combined with meteorological data yield practical oversizing results for marginal quality rock. Critical assumptions of this methodology are: (1) linearity of weight loss as a function of the number of cycles of freezing and thawing, (2) a semi-arid to arid environment which reduces the influence of chemical weathering, and

Table 2. Average percent weight loss for riprap samples during standard durability tests.[1]

Rock Sample	No. of Samples[2]	Na_2SO_4 Soundness (5 cycles)	LA Abrasion (100 cycles)	Freezing and Thawing (250 cycles)
Quarried, unweathered Wyoming granitic rock	5-7 Composite Samples	1.3	5.1	0.1 (0.4×10^{-3})
Granitic boulder conglomerate from a mine	4	5.1	8.8	0.1 (0.4×10^{-3})
Granitic boulder conglomerate from an outcrop	4	4.0	7.2	0.1 (0.4×10^{-3})
Calcite cemented sandstone concretions from a mine	4	97.4	27.5	1.4 (5.6×10^{-3})
Calcite cemented sandstone concretions from an outcrop	4	Not tested[3]	Not tested[3]	6.9 (28×10^{-3})
Calcite cemented sandstone concretions from an outcrop. Minor iron oxide cement	4	99.7	42.3	0.7 (2.8×10^{-3})

1 Tests were performed by the U.S. Bureau of Reclamation (USBR), % weight loss per cycle in parentheses.

2 No. of individual samples unless otherwise noted.

3 According to the USBR, sample would probably have completely disintegrated had this test been performed.

Table 3. U.S. Bureau of Reclamation standards for judging riprap durability. [4]

Test	Quality		
	Poor	Intermediate	Good
Bulk specific gravity	<2.5	2.5 to 2.65	>2.65
Absorption (% weight gain)	>1.0	0.5 to 1.0	<0.5
Freeze-thaw weight loss, %[a]	>5	0.5 to 5	0 to 0.5
Na_2SO_4 weight loss, %[b]	>10	5 to 10	<5
Los Angeles abrasion weight loss, %[c]	>10	5 to 10	<5

[a] 250 cycles
[b] 5 cycles
[c] 100 revolutions

(3) areas to be protected by riprap are well drained. The accuracy of oversizing estimates also depends on how well one can predict the annual number of effective freezing and thawing cycles that occur while riprap is nearly or completely saturated.

Some mill operators and their contractors object to the use of freezing and thawing tests because of their long duration and cost. One might be tempted to develop the Los Angeles abrasion test as a substitute for the freezing and thawing test, based on the fair degree of correlation that apparently exists between the two tests as shown in Table 3. However, marginal quality rock did not perform very well in Los Angeles abrasion tests relative to freezing and thawing tests. Because of its potential use in oversizing marginal quality rock the freezing and thawing test should be retained for evaluating the durability of riprap. However, the number of cycles of such tests might be reduced while still providing useful data for the estimation of oversizing.

ACKNOWLEDGMENTS

This research is sponsored by the Nuclear Regulatory Commission, Division of Waste Management, Office of Nuclear Material Safety and Safeguards under contract with Martin Marietta Energy Systems, Inc.

Special recognition is given to R. H. Ketelle of Oak Ridge National Laboratory for his unselfish support in the preparation of this paper and to R. L. Volpe (R. L. Volpe and Associates) for his participation in sample collection.

Table 4. Oversizing ratios (r_o/r_d) for 200 and 1000 year performance periods and annual, 5-year and 100-year flood recurrence intervals.

Rock Sample	Flood Recurrence Interval: Annual, Performance Period 200 years	Annual, 1000 years	5-year, Performance Period 200 years	5-year, 1000 years	100-year, Performance Period 200 years	100-year, 1000 years
Quarried, unweathered Wyoming granitic rock	1.01	1.06	1.00	1.01	1.00	1.00
Granitic boulder conglomerate from a mine	1.01	1.06	1.00	1.01	1.00	1.00
Granitic boulder conglomerate from an outcrop	1.01	1.06	1.00	1.01	1.00	1.00
Calcite cemented sand stone concretions from a mine	1.22	Unacceptable (2.13 - 400 yrs)	1.03	1.22	1.01	1.01
Calcite cemented sandstone concretions from an outcrop	Unacceptable (1.31 - 50 yrs)	Unacceptable	1.21	Unacceptable (2.05 - 400 yrs)	1.01	1.04
Calcite cemented sandstone concretions from an outcrop. Minor iron oxide cement	1.09	Unacceptable (1.22 - 400 yrs)	1.02	1.09	1.00	1.00

REFERENCES

1. Wardwell, R. E., J. D. Nelson, S. R. Abt, and W. P. Staub. "In Situ Dewatering Techniques for Uranium Mill Tailings," USNRC Report ORNL/TM-8689 (1983).

2. "Environmental Standards for Uranium and Thorium Mill Tailings at Licensed Commercial Processing Sites," Federal Register, Vol. 48, No. 196 (Friday, October 7, 1983).

3. Foley, M. G., C. S. Kimball, D. A. Myers and J. M. Doseburg. "The Selection and Testing of Rock for Armoring Uranium Tailings Impoundments," USNRC Report PNL-5064 (1985).

4. Nelson, J. D., S. R. Abt, R. L. Volpe, D. Van Zyl, N. E. Hinkle and W. P. Staub. "Methodologies for Evaluating Long-Term Stabilization Designs of Uranium Mill Tailings Impoundments," USNRC Report ORNL/TM-10067 (1986).

5. U.S. Bureau of Reclamation, Unpublished open-file data (1946-1985).

6. "Riprap Quality Evaluation," U.S. Bureau of Reclamation. Unpublished Report for the U.S. Department of Energy (1986).

DESIGN OF SOLID WASTE DUMPS AND WASTE DISPOSAL IN WESTERN CANADIAN SURFACE MINES

R. K. Singhal, Group Leader, CANMET, Surface Mining Laboratory, Devon, Alberta, Canada

T. Vladut, President, Retom 1985 Research & Development Ltd., Calgary, Alberta, Canada

ABSTRACT

A sizeable surface mining industry exists in Western Canada, including surface coal mines in Alberta, British Columbia and Saskatchewan, the oilsands mines in Fort McMurray and many the metal mines in British Columbia, Yukon and the Northwest Territories. This paper describes the geotechnical issues involved in designing, contructing and maintaining the stability of solid waste dumps in Western Canadian surface mines.

INTRODUCTION

The bulk of the mineral production in Western Canada is from surface mines. Fifty-seven million metric tonnes (57×10^6 t) of coal, from a total Canadian production of sixty million tonnes (60×10^6 t), is produced in Western Canada by surface mining methods. The production of this tonnage requires the handling of an estimated 200×10^6 m^3 of overburden materials per year. The two oilsands plants operating in Alberta, which produce 200,000 bbls of synthetic crude oil per day, are surface mining operations. These operations at present require the handling of an estimated 25×10^6 m^3 of overburden and solid reject material per annum. The metalliferous mining industry adds another estimated 30 x $10^6 m^3$ of solid waste. Thus total waste handling in western surface mining, as a conservative estimate, is 255 x $10^6 m^3$/annum. This is an enormous volume of waste material. The selection of specific disposal sites through careful evaluation of geotechnical and economic parameters is an equally enormous task.

SITING OF WASTE DUMPS

The most important consideration in the siting of waste dumps is economics; that is to say that the material must be disposed as close to the pit boundary as possible to minimize haulage cost, with due regard to the mine planning requirements of ensuring the required supply of ore now and in the future as the pit plans dictate. There are also exceptional site-specific situations where not-so-standard concepts apply, and waste must be disposed not above the ground but below it as in backfilled pits, valleys and underground disposal facilities.

Optimum location of waste dump sites is first determined from the short-term and long-term mine plans. When the potential sites have been thus selected, a geotechnical investigation is then conducted to determine the suitability of sites. The geotechnical investigation will normally include engineering and geological site inspection, foundation and construction, soil sampling and analysis, properties of materials comprising waste, topography and the proposed method of waste material transportation. The foundation characteristics which affect the siting and stability of waste dumps are: shear strength, compressibility and permeability. The material characteristics which influence the design, geometry and stability of waste dumps are: gradation (grain size and distribution), density, permeability, shear strength, moisture content and plasticity. Changes in waste material properties due to weathering and water must also be considered.

CONSTRUCTION OF WASTE DUMPS

In Western Canada, the most common method of transporting waste to dump sites is by truck. Figure 1 illustrates two methods of dumping. In the "end-tipping" method (Figure 1a) the material is simply discharged to compact under its own weight. This is the common practice in many of the surface mines in mountainous regions. Figure 1b illustrates that method of dumping in layers, also known as "area dumping". As truck loads are dumped on a prepared area, they are spread by a dozer. Area dumping then continues on the next elevation and so on. This dumping method has the advantage that the traversing truck traffic continues to compact the dumped material, improving dump stability. This method of dump construction has been particularly useful at oilsands operations for building overburden dykes to contain tailings. There, rigid construction requirements demand dumping in layers few inches at a time, with each layer being suitably compacted.

Figure 2 shows the four types of dumps found in mountain coal mines. Free dumps are typically formed by end-tipping. Wrap-around dumps provide stability to high dumps. The lifts are usually 25 to 40 m high, with the exact height usually decided by practical considerations of pit ingress at lower elevations. Toe-berm dumps provide stability to high dumps by reinforcing the toe. Built-up dumps, like free dumps, are usually formed by end-tipping. These dumps are built in successive layers: layers 2 and 3 will be built by a combination of end-tipping and area dumping, providing overall greater dump stability. In this way, within the same real estate, more waste can be accomodated.

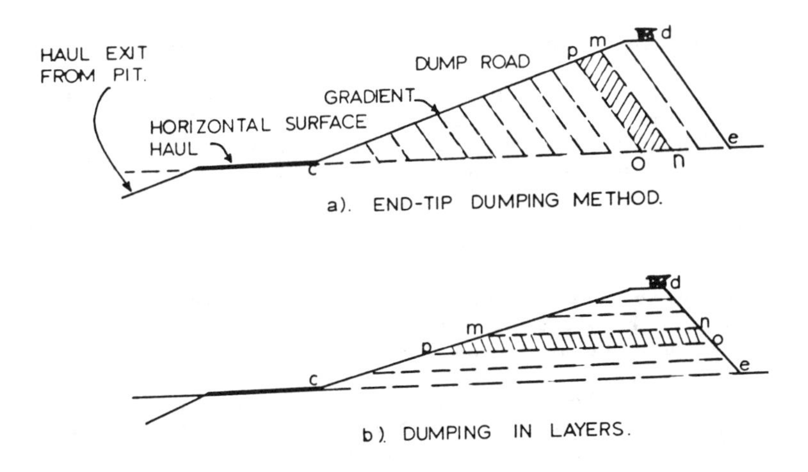

Figure 1. Truck dumping Methods (Ref 1).

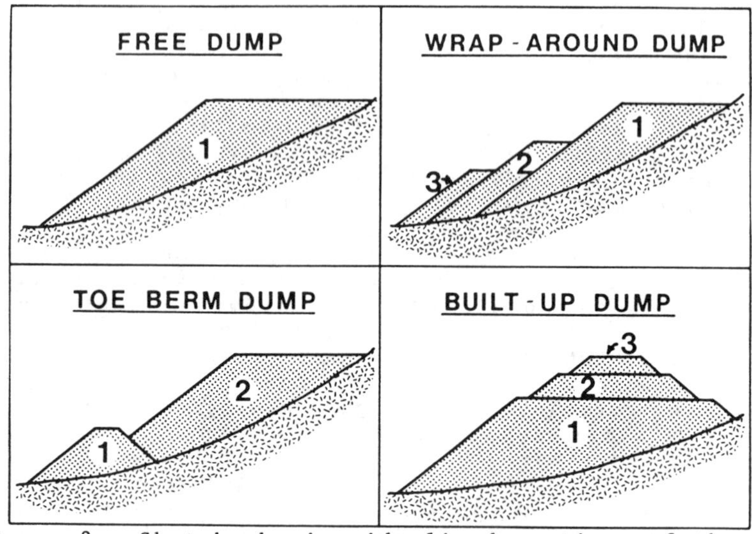

Figure 2. Sketch showing idealized sections of the most common waste rock dump types in mountain coal mines. The numbers indicate sequence of development (Ref 5).

WASTE DUMP STABILITY

Most waste dump failures result from weak foundations, placement of waste materials on too steep a slope, or too great a height and high piezometric water levels. Dump failures usually occur through sliding, and therefore resistance to sliding along potential surfaces within the dump and its foundation is a major factor in the stability of a waste dump. This resistance is governed by the shear strength of the materials and the pore water pressure at the failure surface.

Weathering and softening by water can reduce the shear strength of some materials. Materials such as hard rock, gravel and sand are affected slowly by chemical action, weathering and the passage of time, whereas others such as clays, shales, and siltstones break down rather quickly. Water pressures will vary within the waste dump and its foundation depending on the source of seepage water and relative permeability of the various materials comprising the waste dump. Changes in water pressure (pore pressure) will affect the resistance of a waste dump to sliding; it is one parameter which should be carefully monitored, particularly in problem areas, by providing proper water management: preventing surface water from seeping through the waste piles, clearing snow accumulation quickly and preventing blockage in drainage culverts beneath and around the waste dumps.

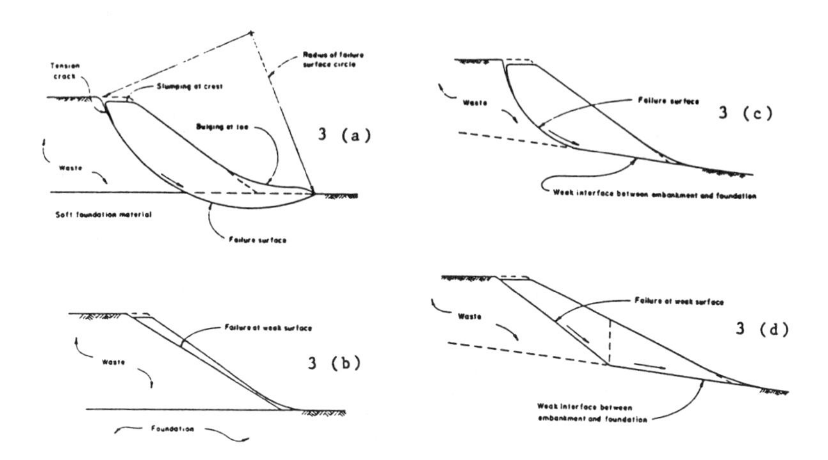

Figure 3. Typical instability modes: a) rotational sliding; b) plane shear; c) non-circular sliding; d) wedge instability (Ref 2).

Figure 3 shows the four types of instability modes encountered in dump failures. In rotational sliding (Figure 3a), the surface along which movement occurs approximates a segment of a horizontal cylinder or sphere. Such slips are indicated by cracking, by tension cracks at the top of the failure surface, by surface depression at the top of the slip surface and by up-thrust and bulging at the toe. Figure 3b depicts a special case of rotational slide where the radius of the failure surface is large and the failure surface almost planar. This type of instability is manifested in waste dumps containing dry cohesionless materials, formed by dumping at the natural angle of repose of the material. Non-circular sliding and wedge failure, Figure 3c and 3d, occur in situations where a weak surface exists beneath the slope. Such a situation arises when layers of fine material are included in a pile of coarse waste and/or when waste material has been piled on a snow covered surface.

The other types of instability and failure modes are creep, flow slide and mud run. Creep in effect is a slow rotational slip of a very large radius and is accompanied by widespread cracking, shallow rotational slips and bulging at the toe. As the name implies, in this case the material moves down at a slow and steady rate parallel to the slope. In a flow slide, a mass of very loose material may behave as a liquid if subjected to large strains; large movement of rotational slip can develop into a flow slide. In mud run conditions, finely divided material and water move under gravity in the form of a viscous fluid. Such a situation is seen in case of waste dumps constructed of relatively fine and impervious material, constructed by dumping over the side on steep slopes.

WASTE DUMP CONSTRUCTION IN SURFACE MINING OF OILSANDS

The soil types encountered in oilsands mining include peat-muskeg, sand-alluvial and lacustrine, gravel-alluvial, clay, clay (till), clay and sand (Clearwater Formation) and oilsands. Most of these materials have low strengths. Their permeability is also low; some materials are practically impervious and as a result, water does not drain easily. Silty materials breakdown completely in the presence of water, becoming sludge. Swelling of wetted clay in clay-rich strata of low density is a cause for concern in the stability of waste dumps and dykes constructed to retain tailings.

In oilsands mining, waste dumps are designed to provide storage for consolidated overburden, and to contain the muskeg and other soft and wet overburde materials. The basic design concept for waste dumps is to construct a retaining shell of inorganic overburden material to provide an enclosure in which to place the muskeg and other soft inorganic overburden materials that are not suitable for inclusion in the retaining shell (Figure 4). The waste dumps are generally constructed to a height of 30m above the original ground level, with outside side slopes of 3 to 1 (horizontal to vertical), and 9m berms on a 7.6m vertical spacing, providing an overall outside slope of approximately 4 to 1.

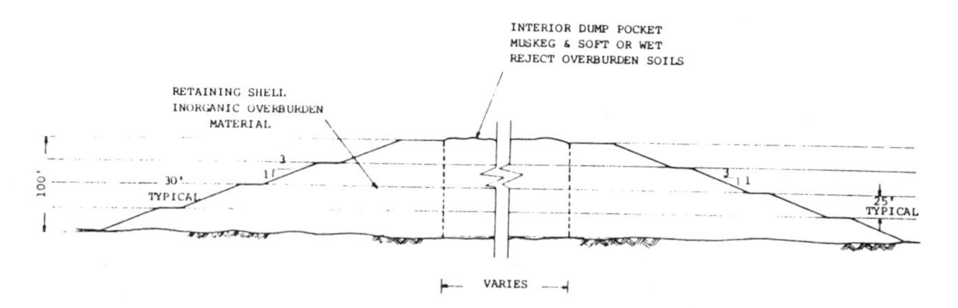

Figure 4. Waste dump typical cross section in an oilsands mine.

The hot water separation process used in oilsands produces large volumes of tailings. These must be retained in tailings ponds situated within, and in some cases, external to the pit. In case of in-pit storage, the dykes must contain the tailings placed in the mined-out areas of the pit without affecting active mining operation. Therefore, dyke side slopes and internal drainage features (zoning) are designed and constructed in accordance with accepted engineering practice for water retention structures.

Figure 5 shows a typical cross-section for an in-pit dyke. Such dykes may reach a height of 90m and are constructed in a configuration which provides a compatible balance between dyke stability conditions and quantities of solid overburden materials available for their construction. Dykes are designed with 2 to 1 (horizontal to vertical) side slopes, and 6.1 to 9.1m wide berms on 13.7m vertical spacings, producing an overall 2.5 to 1 side slope.

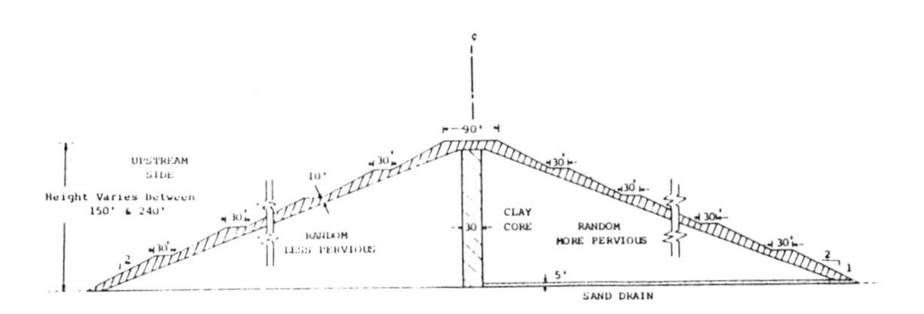

Figure 5. In-pit dyke typical cross section for an oilsands mine.

ENHANCEMENT OF WASTE DUMP STABILITY AND REMEDIAL MEASURES

The design and construction of a waste dump and the selection of a site for its placement are governed by numerous site-specific parameters. For each property, however, there must be guidelines to enable the operating staff to detect and assess the stability of waste dumps and dykes. In addition to visual inspections, which are very meaningful in their own right, various types of instrumentation monitoring can be used. Development of cracks at the crest or other parts of a dump changes its geometry, and signs of foundation material being pushed out should all be noted.

Long-term waste dump stability can be significantly improved by ensuring that water does not find ingress to dumped material. It is the most likely cause of instability since it may lead to both increased pore pressure, and to a softening and breakdown of some materials to form sludge as well as to a reduction in material shear strengths. Therefore, a proper water management program must be implemented to divert the water away from the pit environs and into suitable drains. Drains and culverts must be large enough to carry the maximum amount of water produced including that from storm run-off and spring melting. In some situations, incorporation of geotextiles can contribute to waste dump stability.

The remedial measures to maintain and improve stability of waste dumps are site-specific and will vary from dump to dump. However, typical procedures, most of which are self-explanatory, are shown in Figure 6.

WASTE DUMP RECLAMATION

In the Western Canadian surface mines, particularly coal operations in the mountains, waste dumps are constructed by end-tipping, allowing waste material to settle at slopes of 35 to 37 degrees according to their angle of repose. Once a dump has been completed, it is required to be reclaimed. Regulations in this regard exist almost in all provinces. A similar situation exists for the in-pit waste dumps created by backfilling of mine out areas.

Final dump slope angles will not be greater that 27 degrees. As a minimum 5m wide benches will be maintained between lift of the dump. Erosion control seeding is usually applied to dump surface followed by tree planting on the crest of the waste dump. Figure 7, which reads left to right, illustrates a typical sequence of reclamation.

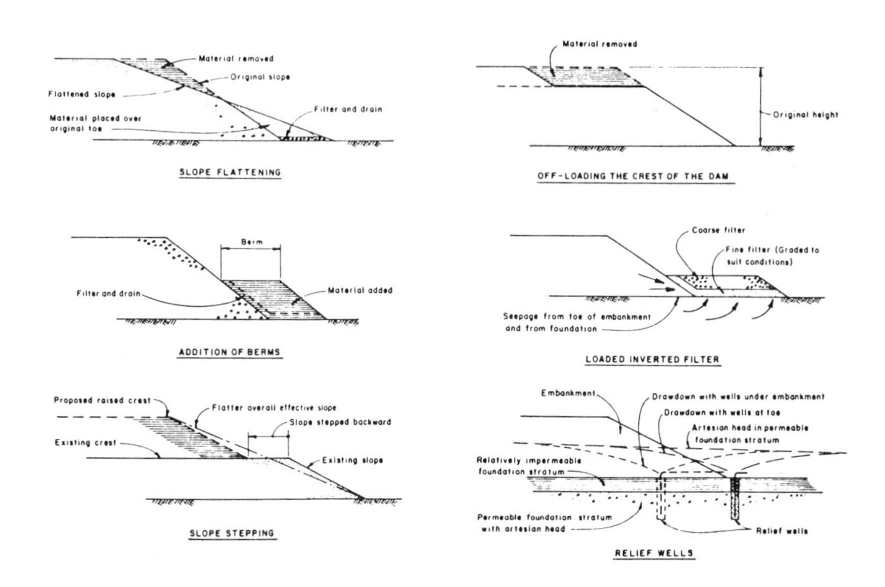

Figure 6. Remedial measures for improving dump stability (Ref 2).

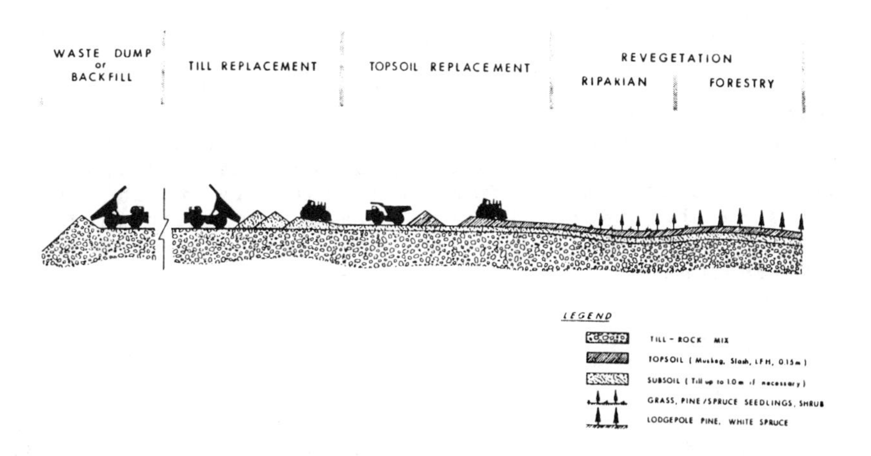

Figure 7. Reclamation of waste dumps (Manalta Coal Ltd.).

GENERAL DISCUSSION

In many Western Canadian mines, particularly those in mountains, end-tipping (end dumping) procedure is followed. The dumping of overburden from the valley wall or the crest of an intermediate dump face results in favourable segregation of material and in the partial compaction of the waste materials. The stability of all waste dumps is primarily dependent on the stability of its foundation materials when loaded by the waste and exposed to internal pressures. It is imperative that, prior to the construction of the dump, vegetation, snow and fine and incompetent material of low shear strength be removed from the foundation surface.

Seepage and pore pressure acting within and below a waste dump should be monitored prior and during the construction of the dumps. Construction of waste dumps often results in changes to existing groundwater conditions including pore pressures. From the viewpoint of stability, the most critical time period for a dump is during and immediately following their construction. With time excess pore pressure gradually dissipate. The importance of reviewing the regional and local hydrogeology to determine typical groundwater conditions within the mine area must be recognized and must be carried out during the pre-planning stage.

In some extreme cases dewatering might be necessary to minimize pore water pressure. In water-saturated unconsolidated overburden special material handling and disposal techniques have been necessary. A significant improvement in the handleability of such overburden can be realized by general pit-dewatering.

Economics of material handling plays a significant role in site selection for waste dumps. The challenge facing the mine planners and the geotechnical engineers is to provide an optimal solution by balancing material handling costs on one hand and stability of the waste dumps on the other. In this regard, the preformance of the Western Canadian surface mine operators has been outstanding.

REFERENCES

1. Steffen, O.K.H., W. Holt, V.R. Symons. "Optimizing Open Pit Geometry and Operational Procedure" in Planning Open Pit Mines, P.W.J. Van Rensburg, ed. (Amsterdam: A.A. Balkema, 1970).

2. Pit Slope Manual, (Ottawa: Canada Centre for Mineral and Energy Technology, 1977).

3. Stepanek, M. "Stability of Waste Embankments in a Northern Environment" in Geotechnical Stability in Surface Mining, Raj K. Singhal, ed. (Rotterdam: A.A. Balkema Publishers, 1986).

4. Singhal, R.K., R.J. Kolada, T.I. Vladut. "Productivity and Equipment Selection in Surface Mining of Oil Sands" in Transactions Society of Mining Engineers, U.S.A., 1986.

5. Claridge, F.B., R.S. Nichols, A.F. Stewart. "Mine Waste Dumps Constructed in Mountain Valleys". CIM Bulletin: 79:87 (1986).

6. Personal Communication: D.S. Cavers (Hardy Associates); Milos Stepanek (Geo-Engineering Ltd.); R.D. Gold (Fording Coal Ltd.); D. Stewart (Afton Operating Corporation).

GEOCHEMICAL RESULTS OF LEACHING SHALE AT AMBIENT TEMPERATURE AND 100°C

Kathryn O. Johnson,
GECR Inc., Rapid City, South Dakota

INTRODUCTION

Oak Ridge National Laboratory is conducting research for the U.S. DOE to evaluate the potential for sedimentary rocks to serve as a repository for high level nuclear waste [1]. The work reported in this paper is part of these efforts and represents an attempt to characterize the geochemical behavior of various shale types at high temperatures. Experiments were designed to analyze the mineralogical alterations and the changes in the pore water chemistry which occur when the temperature is increased.

This study centered on the batch leaching of different shale types for several weeks in distilled water at ambient temperature and at 100°C. The composition of the water after leaching and the resulting mineralogy were determined. These data were used for mineral equilibria calculations by the thermodynamic computer code, PHREEQE. The geochemical reactions that occur with a temperature increase and the mineral equilibria that control the aqueous chemistry were inferred from the relationships in the analytical data set and the results from PHREEQE.

Shales were chosen to represent the various compositional parameters thought to be important in the geochemical alterations of shales at high temperature: clay mineralogy, calcite, pyrite, and organic material. Swelling smectite clays, such as montmorillonite, undergo alterations at high temperature to illite and chlorite [2]. This is accompanied by loss of water and cation exchange reactions that create changes is physical and chemical properties. Calcite is important as a buffer against pH changes by acid produced from reactions that may occur at high temperatures. Pyrite is easily oxidized at high temperatures, thus representing a source of acid. In addition, pyrite may serve as a buffer against oxidizing conditions if the pH remains near neutral.

METHODS

The experiments were done on shales from five formations: Pierre, Green River, Conasauga, Chattanooga, and Rhinestreet, chosen to represent the variations in mineralogy within the broad class of shale rock [3]. The Pierre

Geotechnical and Geohydrological Aspects of Waste Management, D. J. A. van Zyl et al., Eds., © 1987 Lewis Publishers, Inc., Chelsea, Michigan – Printed in USA.

Shale of Late Cretaceous Age in South Dakota contains mostly smectite-type clay minerals. In contrast, the clay mineralogy of the Paleozoic Shales: the Devonian Chattanooga and the Cambrian Conasauga from Tennessee, and the Devonian Rhinestreet from West Virginia, is dominated by illite. The Tertiary Green River Shale from Colorado is more aptly called a marl rather than a shale because of the dominance of carbonates and the minimal amount of clay minerals. Samples from several members of the Pierre Shale (Degray, Elk Butte, Mobridge, Verendrye and Virgin Creek) with different mineralogy were available and therefore were included in the experiments.

The leaching experiments were done with 25 g of sample ground to less than 0.15 mm in a volume of 100 ml of distilled water. Nitrogen gas was used to purge the oxygen from the system. Polyethylene and teflon bottles were used for the ambient temperture and 100°C leachings, respectively. The ambient leaching was continued for 43 days, whereas the duration of the leaching at 100°C was 38 days. The slurries of sample and water were agitated for about 30 minutes per day.

ANALYTICAL DATA

The pre-leaching and post-leaching mineralogy of the shale samples are given in Tables 1, 2, and 3. The quartz content is an indicator of the homogeneity of the bulk sample used for the three sets of mineralogical analysis. The quartz content of the pre- and post-leaching samples vary significantly and, therefore, comparisons before and after leaching cannot be made. The percent quartz in the samples used for the low and high temperature leaching are similar in all samples except for the Degray, Elk Butte 1, and Rhinestreet. Because of the apparent heterogeneity in the mineral distributions these samples were not used for comparisons between leaching at low and high temperatures.

The samples from the Pierre Shale contained about 20 weight percent water as determined by drying at 105°C for 24 hours. The other shale types contained no greater than one percent moisture.

The mineralogy shows variability among the members of the Pierre Shale. The sample from the Mobridge 2 appears to be a calcite concretion because of the high calcite and low silicate content. The other members have about the same percentage of feldspar with variable amounts of quartz and clay minerals. The clay mineralogy of the Pierre Shale is dominated by montmorillonite and kaolinite; however, the ratio, (chlorite + illite)/(kaolinite + montmorillonite) varies between 0.2 and 0.8. The Conasauga and the Rhinestreet Shales are dominantly chlorite and illite; whereas, the Chattanooga Shale contains equal amounts of illite and montmorillonite and a high quartz content.

All of the shales except for the Green River contain pyrite which indicates their reducing nature. The carbonate content of the shales is varied. The Pierre Shale is generally carbonate rich except for the Virgin Creek and Degrey members. The Paleozoic Shales are low in carbonates except for the Rhinestreet, which contains over 25 percent carbonates as calcite and siderite. Gypsum is present only in small quantities or is absent; however, the bulk chemical analysis (Table 4) showed higher sulfate concentrations than are accounted for by XRD analysis of gypsum. This discrepancy may be due to the difficulty in quantitative analysis of gypsum by XRD or by an increase in gypsum from the oxidation of pyrite.

The compositions of the resulting leachates are contained in Tables 5 and 6. The data have been standardized as $\mu g/g$ of solid sample. The relationship between the data as $\mu g/g$ of solid sample and $\mu g/ml$ of pore water depends upon the bulk density, the porosity, and the water saturation of

Table 1
Mineralogy of Shale, Weight Percent

MINERAL	SHALE TYPE											
	DeGrey	Elk Butte 1	Elk Butte 2	Mobridge 1	Mobridge 2	Verendrye	Virgin Creek 1	Virgin Creek 2	Green River	Conasauga	Rhinestreet	Chattanooga
Chlorite	1.47	—	—	—	2.27	—	—	—	—	20.40	20.11	—
Illite	5.50	14.25	10.86	13.60	3.49	27.30	34.51	11.99	3.88	53.36	31.02	20.84
Kaolinite	—	7.66	4.13	8.69	1.89	2.05	0.67	1.49	—	—	—	1.40
Montmorillonite	31.15	26.33	31.70	25.10	13.12	32.64	41.05	41.73	—	—	—	20.29
Oligoclase	4.86	4.46	5.26	3.94	2.28	4.90	3.63	3.40	7.92	3.51	—	4.61
Orthoclase	4.28	4.56	4.84	4.11	1.75	4.50	4.78	3.78	5.86	—	—	5.81
Quartz	45.47	23.61	26.32	24.86	10.56	25.47	13.99	34.87	17.52	20.53	20.56	40.51
Analcite	—	—	—	—	—	—	—	—	8.32	—	—	—
Stilbite	—	—	3.82	—	—	—	—	—	—	—	—	—
Calcite	—	16.84	9.05	18.57	56.45	1.73	—	—	6.40	—	5.97	—
Dolomite	—	—	—	—	—	—	—	—	44.10	—	—	—
Siderite	—	—	1.21	0.48	—	—	0.31	0.47	—	—	19.45	—
Gypsum	1.06	—	1.15	—	2.47	—	—	—	—	—	—	—
Bassanite	—	1.51	—	—	—	0.63	—	1.18	—	—	—	—
Pyrite	1.13	0.72	0.90	0.59	1.05	0.74	1.02	1.05	—	2.18	0.43	6.50
Jarosite	—	—	0.71	—	—	—	—	—	—	—	—	—
Glauconite	—	—	—	—	—	—	—	—	—	—	—	—
Limonite	5.05	—	—	—	—	—	—	—	—	—	—	—

— None detected by XRD

Table 2
Mineralogy of Shale After Leaching at Ambient Temperature, Weight Percent

MINERAL	SHALE TYPE										
	DeGrey	Elk Butte 1	Elk Butte 2	Mobridge 1	Mobridge 2	Verendrye	Virgin Creek 1	Virgin Creek 2	Green River	Conasauga	Rhinestreet	Chattanooga
Chlorite	1.73	—	—	—	—	—	—	—	—	17.81	20.14	—
Illite	6.37	6.93	10.25	5.90	2.05	8.84	6.65	8.75	1.65	30.70	27.66	22.84
Kaolinite	—	6.92	3.64	7.43	2.35	2.74	—	3.75	—	—	—	1.40
Montmorillonite	25.19	28.36	26.34	30.56	16.33	34.14	55.23	29.01	3.23	26.87	—	22.40
Oligoclase	6.33	5.11	6.81	5.34	2.68	5.28	2.38	1.87	9.10	3.54	—	5.09
Orthoclase	5.18	3.02	4.07	3.65	1.72	4.97	—	3.89	5.11	0.55	2.44	6.78
Quartz	45.21	31.64	27.49	29.70	15.48	40.53	33.33	49.90	19.04	14.46	22.05	34.42
Analcite	—	—	—	—	—	—	—	—	9.07	—	—	—
Stilbite	1.87	—	8.65	—	—	—	—	—	—	—	—	—
Calcite	—	16.92	9.63	16.06	57.06	1.26	—	—	6.39	—	6.13	—
Dolomite	1.60	—	—	—	—	—	—	—	46.37	—	—	—
Siderite	—	—	1.16	0.50	—	—	—	—	—	—	21.27	—
Gypsum	—	—	—	—	0.43	—	—	—	—	—	—	—
Bassanite	—	—	—	—	—	1.14	—	—	—	—	—	—
Pyrite	0.85	1.07	1.27	0.81	1.85	1.04	2.37	2.79	—	1.06	0.28	7.03
Jarosite	—	—	0.64	—	—	—	—	—	—	—	—	—
Glauconite	—	—	—	—	—	—	—	—	—	5.00	—	—
Limonite	5.62	—	—	—	—	—	—	—	—	—	—	—
Aluminosilicate Ratio [(a)]	0.32	0.20	0.34	0.16	0.11	0.24	0.12	0.26	0.51	1.81	∞	1.02

— None detected by XRD

(*a*)(Chlorite + illite)/(kaolinite + montmorillonite)

Table 3
Mineralogy of Shale After Leaching at 100°C, Weight Percent

MINERAL	SHALE TYPE											
	DeGrey	Elk Butte 1	Elk Butte 2	Mobridge 1	Mobridge 2	Verendrye	Virgin Creek 1	Virgin Creek 2	Green River	Conasauga	Rhinestreet	Chattanooga
Chlorite	1.09	—	—	—	—	—	—	—	—	3.20	1.60	—
Illite	11.83	6.16	10.58	6.52	3.18	7.44	7.80	7.52	2.52	25.60	24.44	43.43
Kaolinite	—	7.65	5.04	6.66	1.95	2.90	—	—	—	4.01	15.01	—
Montmorillonite	36.12	24.23	35.84	31.49	19.45	37.33	50.90	33.46	3.73	36.53	—	—
Oligoclase	5.84	4.38	7.93	6.47	3.00	6.29	8.18	4.57	7.68	3.05	4.39	7.69
Orthoclase	3.80	2.78	4.64	3.77	1.99	4.04	—	3.47	9.74	0.31	2.61	5.97
Quartz	39.44	38.12	26.24	33.29	13.46	37.16	32.57	50.11	16.60	14.93	40.96	37.08
Analcite	—	—	—	—	—	—	—	—	7.48	—	—	—
Stilbite	—	—	—	—	—	—	—	—	—	—	—	—
Calcite	—	13.25	5.37	9.30	54.42	0.54	—	—	6.22	—	—	—
Dolomite	1.85	—	—	—	—	—	—	—	45.99	—	—	—
Siderite	—	—	0.65	—	—	0.66	—	—	—	—	1.59	—
Gypsum	—	—	0.40	—	1.74	—	—	—	—	—	—	—
Bassanite	—	—	—	—	—	—	—	—	—	—	—	—
Pyrite	—	0.78	0.41	0.53	0.75	—	0.53	0.84	—	1.21	—	5.52
Jarosite	—	—	0.25	—	—	—	—	—	—	—	—	0.28
Glauconite	—	—	—	—	—	—	—	—	—	0.83	5.27	—
Limonite	—	2.60	2.60	—	—	3.60	—	—	—	4.29	4.07	—
Aluminosilicate Ratio [a]	0.36	0.19	0.26	0.17	0.15	0.18	0.15	0.22	0.69	0.71	1.73	∞

— None detected by XRD
(*a*)(Chlorite + illite)/(kaolinite + montmorillonite)

Table 4
Chemical Composition of Shale, $\mu g/g$

ELEMENT	SHALE TYPE											
	DeGrey	Elk Butte 1	Elk Butte 2	Mobridge 1	Mobridge 2	Verendrye	Virgin Creek 1	Virgin Creek 2	Green River	Conasauga	Rhinestreet	Chattanooga
CO_3	4.16	6.84	5.08	18.72	4.20	4.40	0.92	2.76	16.0	3.30	4.32	3.36
SO_4	0.33	0.56	1.15	0.61	2.53	0.39	1.38	1.71	0.15	0.72	0.53	0.43
Ca	1.43	4.35	3.51	5.01	17.8	2.02	2.20	1.56	12.2	1.10	0.47	0.26
Mg	0.05	0.08	0.79	0.16	0.003	0.11	6.53	2.77	0.13	0.29	0.21	0.007
K	0.25	0.62	0.57	0.50	0.40	0.34	1.00	1.25	0.63	0.68	0.59	0.55
Na	0.29	2.40	1.05	0.97	0.001	0.55	2.90	3.10	1.90	1.30	1.50	1.10
Al	7.23	8.70	8.48	8.03	4.75	8.48	8.48	7.23	4.78	13.0	10.01	6.45
Fe	3.05	3.95	4.38	3.65	4.58	3.70	3.05	3.93	2.03	6.10	5.63	4.33
Mn	0.51	0.007	0.048	0.04	0.002	0.34	0.011	0.15	0.02	0.01	0.06	0.001
SiO_2	61.4	55.8	60.1	56.9	35.6	77.4	58.6	63.5	46.3	69.5	56.0	65.7
S[a]	3.30	5.70	5.89	7.52	3.38	4.12	3.07	4.32	4.12	4.93	5.76	7.58
TOC[b]	1.89	0.95	2.15	0.66	3.79	1.02	0.86	0.81	9.09	0.62	0.60	6.27

(*a*)Total sulfur minus sulfate
(*b*)Total organic carbon

Table 5
Chemical Composition of Water Leachate at Ambient Temperature, $\mu g/g$

ELEMENT	SHALE TYPE											
	DeGrey	Elk Butte 1	Elk Butte 2	Mobridge 1	Mobridge 2	Verendrye	Virgin Creek 1	Virgin Creek 2	Green River	Conasauga	Rhinestreet	Chattanooga
HCO_3	1,700	1,080	1,020	1,200	1,420	1,340	420	980	2,240	1,120	1,260	1,220
Cl	300	100	140	200	200	240	200	220	140	800	240	380
SO_4	3,560	7,280	7,800	6,760	12,000	4,360	8,960	10,640	280	380	80	120
Ca	140	1,980	2,120	1,980	4,026	200	570	1,980	170	20	240	140
Mg	40	264	264	264	330	42	150	396	166	62	70	40
K	140	340	260	340	320	160	320	700	60	420	180	240
Na	200	600	740	600	740	200	360	340	400	780	420	640
Al	6	6	6	6	6	8	6	6	6	724	8	8
Fe	1	1	1	1	1	1	4	1	1	82.8	1	1
Mn	2.4	12.8	13.4	13.6	1.8	3	5.2	308	0.6	0.6	0.6	0.6
SiO_2	20	20	20	20	2	26	91.6	91.6	20	1316	26	53
Eh[(a)], mV	-51	-48	-47	-62	-45	-50	57	83	-158.3	-235.467	-122	-14
pH	7.4	7.4	7.4	7.5	7.4	7.4	6.5	6.6	8.2	7.8	7.9	7.1
TDS[(b)]	8,080	12,020	12,740	11,600	19,260	8,700	14,160	17,120	3,600	6,840	2,960	3,360

(*a*)Eh calculated by PHREEQE assuming equilibrium with finely divided crystalline ferric hydroxide

(*b*)Total Dissolved Solids, $\mu g/g$, calculated by PHREEQE

Table 6

Chemical Composition of Water Leachate at 100°C, μg/g

ELEMENT	SHALE TYPE											
	DeGrey	Elk Butte 1	Elk Butte 2	Mobridge 1	Mobridge 2	Verendrye	Virgin Creek 1	Virgin Creek 2	Green River	Conasauga	Rhinestreet	Chattanooga
HCO_3	1,260	1,460	1,260	1,220	1,760	1,220	40	260	40	200	1,360	40
Cl	120	80	110	90	120	80	56	60	80	880	140	220
SO_4	7,820	9,020	8,660	7,460	11,880	7,860	11,360	14,960	2,760	9,320	6,640	11,720
Ca	1,340	3,320	3,080	2,820	4,200	1,840	1,160	1,980	340	640	2,260	2,040
Mg	180	440	240	260	240	160	440	740	80	1,260	500	600
K	480	520	720	440	340	480	520	1,040	100	920	360	720
Na	2,700	760	960	760	800	2,820	2,880	3,040	1,480	1,800	640	800
Al	11	14	3.8	5.8	2	3.8	34	20	2	14	2	86
Fe	2.2	1.8	1	1	1	1.8	312	110	1	48	1	1,604
Mn	131	12.4	14.2	6	0.6	30.6	11.8	464	0.6	58	13.6	32.2
SiO_2	174	81.6	81.6	81.6	66.2	132	522	270	114	270	66.2	522
$Eh^{(a)}$, mV	-206	-260	-241	-257	-271	-230	365	160	-274	157	-242	310
pH	7.1	7.5	7.5	7.6	7.7	7.3	3.4	4.5	7.7	4.6	7.5	3.4
$TDS^{(b)}$	15,460	17,020	16,280	14,220	19,600	16,720	18,380	24,140	6,200	16,440	13,000	20,380

(*a*)Eh calculated by PHREEQE assuming equilibrium with finely divided crystalline ferric hydroxide

(*b*)Total Dissolved Solids, μg/g, calculated by PHREEQE

the pore space. The major chemistry of the leachates of the Pierre Shale is dominated by calcium and sulfate. In contrast, the major ionic constituents in the leachates from the Paleozoic Shales are sodium and chloride.

The relationship between the paucity of calcite in some of the shales and the acidic conditions of the high temperature leachates demonstrates the role of calcite in buffering the pH of the pore water. At high temperatures the oxidation of pyrite occurs at an accelerated rate. This generates acid which if not neutralized by the dissolution of carbonate minerals lowers the pH. The acidic conditions allow for the high concentrations of aluminum and iron in the leachate.

The Eh values were calculated by PHREEQE assuming equilibrium with finely divided crystalline ferric hydroxide. The only significance of these values is that the reducing environment appears to have been maintained through the low temperature leaching process. The calculated Eh values for the high temperature leaching suggests that in some cases the redox potential became oxidizing during the leaching process. The pH of the leachates remained near neutral except for a few of the high temperature leachates which became acidic.

The TDS of the leachates at low temperature are consistently lower than the TDS of the leachates at high temperatures. The increase in TDS at the high temperature is greater in the Paleozoic Shales than in the Pierre Shale. This suggests that the effects of high temperature on chemical alterations is greater in older shales than in younger shales.

MINERAL EQUILIBRIA

Mineral equilibrium calculations on the extracts from the batch leaching were done with PHREEQE [4]. PHREEQE expresses the degree of thermodynamic mineral saturation with a saturation index (SI). A SI is the logarithm of the ratio of the activity product (IAP) of the component ions of the solid in solution to the equilibrium solubility product (Keq) of the solid. A SI of zero indicates equilibrium, a SI of less than zero denotes undersaturation (dissolution), and a SI of greater than zero denotes supersaturation (precipitation).

The saturation indexes for gypsum, calcite, and halite are given in Table 7. The SI values of halite show that it is consistently undersaturated. This indicates that the shales did not contain enough halite to maintain chemical saturation in 100 ml of water. This does not provide a conclusion as to the degree of halite saturation in the pore water.

The saturation indexes of calcite agree very well with the mineralogy of shale samples. The SI values that are less than zero (undersaturated) correspond to those samples that contain little or no calcite; whereas, the SI values of greater than zero correspond to those shales that contain calcite. The presence of calcite provides a buffer to maintain a pH of near 7. The shales that lack calcite became acidic during the high temperature leaching, except for the Verendrye member which is buffered by dolomite and the Degrey sample in which case the buffering mechanism is not obvious. The saturation indexes of calcite in the leachates at high temperature that became acidic are less than zero indicating the undersaturated conditions of the solution with respect to calcite.

The correspondence between the saturation indexes of gypsum and the gypsum reported for the mineralogy is not as good as for calcite. This is inferred to be because of the inaccuracies of gypsum analysis by XRD or an elevated sulfate concentration due to the pyrite oxidation during the leaching. The saturation indexes of gypsum in both high and low temperature

Table 7
Saturation Indexes of Gypsum, Calcite, and Halite in Leachates at Ambient Temperature and 100°C

MINERAL		SHALE TYPE											
		DeGrey	Elk Butte 1	Elk Butte 2	Mobridge 1	Mobridge 2	Verendrye	Virgin Creek 1	Virgin Creek 2	Green River	Conasauga	Rhinestreet	Chattanooga
AMBIENT TEMP.	Gypsum	-1.1	0.1	0.1	0.1	0.4	-0.9	-0.4	-0.2	-1.9	-2.7	-2.2	-2.3
	Calcite	-0.3	0.5	0.5	0.7	0.8	-0.3	-1.5	-0.5	0.9	-0.7	0.6	-0.5
	Halite	-6.0	-6.9	-6.9	-6.7	-6.6	-6.1	-6.1	-6.2	-6.9	-5.9	-6.3	-6.0
100°C	Gypsum	-0.1	0.3	0.3	0.2	0.5	0.0	-0.1	0.2	-0.8	-0.4	0.1	-0.1
	Calcite	0.1	0.9	0.8	0.9	1.2	0.4	-9.2	-5.6	-1.4	-6.0	0.8	-7.8
	Halite	-5.4	-5.9	-5.8	-6.0	-6.9	-5.1	-5.5	-5.4	-5.7	-5.4	-6.0	-5.9

leachates of the Elk Butte, Mobridge, and Virgin Creek 2 members suggest the presence of gypsum in these shales. In the leachates at high temperatures the SI values are greater than in the leachates at low temperature and are near zero (equilibrium with respect to gypsum) for all of the shale samples. The near equilibrium conditions suggest that the solubility of gypsum controls the sulfate concentration. Subsequent to the oxidation of pyrite and the generation of sulfate, gypsum precipitates to maintain equilibrium with respect to the chemical composition of the ground water.

ALUMINOSILICATE REACTIONS

The behavior of the silicate mineralogy after leaching at high and low temperature differs between the shale types. Alteration of the clay mineralogy with increasing temperature typically results in a conversion from expandable smectite clays (montmorillonite) to illite by the substitution of aluminum for silicon in the tetrahedral layer. This negative charge is balanced by interlayer potassium and exchange of divalent cations (iron and magnesium) for aluminum in the octahedral layer Hower [5]. Accompanying the increase in illite is the reaction between the iron and magnesium, from the montmorillonite, and kaolinite to produce chlorite. The reaction from smectite to illite occurs by the transformation from smectite to mixed layer illite/smectite clays. The proportion of illite in the mixed layer minerals increases as diagenesis progresses.

Evaluation of the mineralogy data from the leaching experiments was done by calculating the ratio, (illite + chlorite)/(montmorillonite + kaolinite) (Tables 2 and 3). Only those shales with similar quartz contents in the samples from the high and low temperature leaching were evaluated. The samples from the Pierre Shale showed very little change in the ratio of the clay minerals between the high and low temperature. This indicates that the silicate mineralogy of the Pierre Shale was not sensitive to the temperature and time conditions of the experiment.

Contrary to conventional wisdom, the data from the high temperature leaching of the Conasauga Shale showed an increase in the percent montmorillonite and kaolinite and a decrease in the percent illite and chlorite. In contrast to the Conasauga, the Chattanooga showed a complete conversion of the montmorillonite and kaolinite to illite.

The consistent behavior of the various members of the Pierre Shale suggests a significance of the results. The meaning of the results from the Conasauga and the Chattanooga are difficult to assess. There is the possibility that inhomogeniety of the clay minerals in the samples used for the two leaching experiments account for the results, or that the XRD analysis did not distinquish between discrete illite and illite/smectite mixed layer clay minerals [6].

REFERENCES

1. Croff, A. G., T. F. Lomenick, R. S. Lowrie, and S. H. Stow. "Evaluation of Five Sedimentary Rocks Other Than Salt for High-Level Waste Repository Siting Purposes," ORNL-6241 (1985), Volumes 1, 2, and 3.

2. Weaver, C. E. "Geothermal Alteration of Clay Minerals and Shales: Diagenesis," ONWI-21 (1979).

3. Gonzales, S., and K. S. Johnson. "Shales and Other Argillaceous Strata in the United States," ORNL/Sub/84-64794/1 (1984).

4. Parkhurst, D. L., D. C. Thorstenson, and L. N. Plummer. "PHREEQE-A Computer Program for Geochemical Calculations," Water Resources Investigation, *U. S. Geological Survey,* pp. 80–96 (1980).

5. Hower, J., E. V. Eslinger, M. E. Hower, and E. A. Perry. "Mechanism of Burial and Metamorphism of Argillaceous Sediment: 1. Mineralogical and Chemical Evidence," *Geological Society of American Bulletin,* Volume 87, pp. 725–737 (1976).

6. Sroden, J., and D. D. Eberl. "Illite," *Reviews in Mineralogy, Volume 13: Micas,* S. W. Bailey, Ed., Mineralogical Society of America, Washington, DC., pp. 495–544 (1984).

PERFORMANCE ALLOCATION - A SYSTEMS APPROACH TO THE CHARACTERIZATION OF HIGH-LEVEL WASTE REPOSITORY SITES

Larry D. Rickertsen
WESTON Technical Support Team, Washington D.C.

INTRODUCTION

The approach to be used to characterize a site for a mined geologic disposal system for high-level nuclear waste is to focus on issues that specifically address the performance objectives. These peformance objectives include the regulatory requirements of 10 CFR Part 60 such as the specified limit to the cumulative release of radionuclides to the accessible environment. Performance allocation is a systematic way of developing and presenting the strategy to resolve these issues that has been selected to define the characterization program.

The steps of the performance allocation process for a specific issue are the following:

- A licensing strategy for the resolution of the issue is developed; that is, the strategy that would be used if only current data and conceptual models were to be considered is prescribed. The most important features of this strategy to be identified are the elements of the mined geologic disposal system that will be explictly considered in this characterization program.

- The measures of performance for these systems elements relevant to the issue are identified.

- Goals for these performance measures are specified such that, in terms of the conceptual models being considered, the goals are consistent with resolving the

Geotechnical and Geohydrological Aspects of Waste Management, D. J. A. van Zyl et al., Eds., © 1987 Lewis Publishers, Inc., Chelsea, Michigan—Printed in USA.

issue. Each goal is expressed as a value and a needed level of confidence in that value. If more than one conceptual model is being considered for a specific element or process, multiple sets of performance measures and goals may need to be specified.

- Parameters needed to evaluate the performance measures are identified. Goals for these parameters are set consistent with the goals for the performance measures. Again, the goal is expressed as a value and a needed level of confidence for that value. Where possible, the existing level of confidence in that value in terms of the expected range of values should also be provided. Unreasonable parameter goals should not be set.

- Studies are identified to obtain the information needed to evaluate the parameters. Study parameters are identified and explicit relationships between these study parameters and the higher-level issue parameters and other study parameters are given. Goals should be set for the study parameters consistent with the higher-level goals wherever appropriate.

The strategy that results from the performance and parameter goals is used to guide the testing program. As information is acquired from the tests and analyses, it can be used in system performance assessments to compare with the overlying performance objectives. These comparisons may suggest that additional testing may be needed. In this case the performance allocation process will be reapplied and a new strategy developed with a new set of performance and parameter goals.

In order to illustrate the performance allocation process, an example is given for one of the important issues that presently confronts the program: will the mined geologic disposal system meet the system performance objective for limiting 10,000-year cumulative release of radionuclides as required by the standards in 10 CFR 60.113 and 40 CFR 191.13? The performance allocation process is applied to a specific site, the Yucca Mountain Site, one of three sites to be characterized for this purpose. It is to be emphasized that the performance measures, parameters and goals presented here are for illustrative purposes only. The actual issue resolution strategies that will finally be chosen depend on a wide range of considerations and may be much different than that given here. Nevertheless the example is based upon specific site characteristics and current conceptual models and is suited for the purposes of illustration.

The example illustrates several techniques that will be used in developing the site characterization plans. First, the example relies heavily on the information and conceptual models used in the Environmental Assessment (EA) report [1]. These models are simple and have received extensive review during the development of the EA. For these reasons these models are likely to be useful in the demonstrations and in resolving the issues. Furthermore, because information is already presented in the EA, the burden of describing the existing information and the assumptions upon which the performance goals are based can be reduced simply by referencing this material.

A second point is the focus on a few, well-defined elements of the system that are chosen because they allow legitimate simplifications of the characterization program. The approach is to avoid reliance on barriers whose favorable characteristics are poorly understood at present. In the case of this example the key elements relied upon are cumulative release from the waste packages, the ground-water travel time, and retardation of radionuclide transport relative to the ground-water movement. In addition to simplifying the characterization program for expected conditions, this approach drastically simplifies the treatment of unanticipated processes and events since all that must really be considered are those factors that can significantly impact these key elements.

AN EXAMPLE OF AN ISSUE RESOLUTION STRATEGY

The known characteristics of the Yucca Mountain site are described in the EA report[1]. The proposed repository horizon is well above the water table and the waste would be emplaced deep within the mountain, thus taking advantage of the known low percolation flux at depth in the unsaturated zone.

In order to develop an issue resolution strategy for the conditions expected for the repository system, it is assumed that radionuclides can be released from the repository and transported to the accessible environment either by ground water percolating down through the repository or by gaseous transport of volatile radionuclides up to the surface. It is assumed that the waste packages fail at some point due to corrosion of the containers and that volatile radionuclides can escape and be transported through the porous, unsaturated rock to the surface. Radionuclides can also be leached from the waste form by ground water that can enter breached waste packages and these leached radionuclides can then be transported by the ground water down to the water table. At the water table the radionuclides are transported down gradient until they reach the accessible environment. This conceptual model and the details of the processes in this model depend upon the characteristics of the repository, the waste form, and the waste packages, upon site characteristics that determine the thermal, fluid, chemical, and radiation conditions in the vicinity of the waste packages, and upon the characteristics of the site that determine the gaseous and fluid transport of the radionuclides.

Licensing Strategy

The preliminary analyses performed for the EA suggest that under expected conditions the regulatory limits on cumulative release of radionuclides in 10,000 years of 40 CFR 191 will be met by a wide margin at the site. These analyses suggest that releases of radionuclides from the waste packages would be very small, well below this standard. Further, the analyses indicate that the transport of radionuclides by water percolating down through the repository will take much longer than 10,000 years to reach the accessible environment, due both to the long ground-water travel time and the significant chemical and physical retardation of radionuclide migration relative to the ground-water movement. Even for transport of gaseous radionuclides, the release of radionuclides to the accessible environment is expected to be insignificant.

These analyses indicate that important elements of the system relative to cumulative release to the accessible environment include:

- the waste package.
- the portion of the Topopah Springs host rock in the unsaturated zone.
- the portion of the Calico Hills in the unsaturated zone.
- the saturated zone, principally the portion of the Calico Hills and Topopah Springs below the water table.

Furthermore the preliminary analyses presented in the EA suggest that the considerations for performance of these system elements relative to cumulative release for the accessible environment include the following factors:

- Cumulative release of radionuclides from the waste packages in 10,000 years.
- Transport of radionuclides through the specified natural barriers to the accessible environment.
- Transport of gaseous radionuclides upward to the surface.

The site characterization program must address the processes that affect these factors. These processes include radionuclide transport within the waste package (e.g., dissolution, leaching, diffusion, and advection), those that affect transport by ground water (e.g., ground-water movement, radiounclide migration), and gaseous transport (e.g., density and pressure gradients and gas dispersion effects).

Performance Measures and Performance Goals

The factors specified provide the principal performance measures for the barriers identified relative to this issue. Goals chosen for the performance measures are listed in Table 1. These goals are the values which, if achieved, would would give

Table 1. Performance Measures and Goals for Issue Resolution.

Performance Measure	Goal	Needed Confidence
Ground-water Transport of Radionuclides		
1. Waste Packages (Primary Barrier)		
Cumulative release of nuclides in 10,000 years	$M_{wp} < 0.1$	Medium
2. Unsaturated Portion of Calico Hills zeolitized unit (Primary Barrier)		
Ground-water Travel Time	$T > 10000$ yr	0.98
Retardation Factor (Pu, Am)	$R \geq 5$	High
3. Unsaturated Portion of Topopah Springs member (Reserve Barrier)		
Ground-water Travel Time	$T \geq 5000$ yr	High
Retardation (Pu, Am, U, Np, C)	$R \geq 10$	High
4. Saturated Zone (Reserve Barrier)		
Ground-water Travel Time	$T \geq 100$ yr	High
Retardation Factor	$R > 500$	High
Gaseous Transport of Volatile Radionuclides		
1. Waste Packages (Primary Barrier)		
Cumulative release of volatiles in 10,000 years	$Q_{gas} \leq 0.1$	Medium
2. Overlying Units (Primary Barrier)		
Ratio of cumulative release of volatiles to inventory of volatiles	Ratio ≤ 0.1	High
3. Shaft Seals (Primary Barrier)		
Ratio of gas transport properties to rock properties	Ratio < 1	High

confidence that the regulatory requirements would be met for expected conditions. The goals are set so that these requirements would be likely to be met with high confidence either by the engineered barriers alone and by the natural barriers alone.

The goal for the performance of the waste packages is that the cumulative release of radionuclides at the boundary of the waste packages in 10,000 years is a factor of 10 below the standard of 40 CFR 191.13. The performance measure in Table 1 is

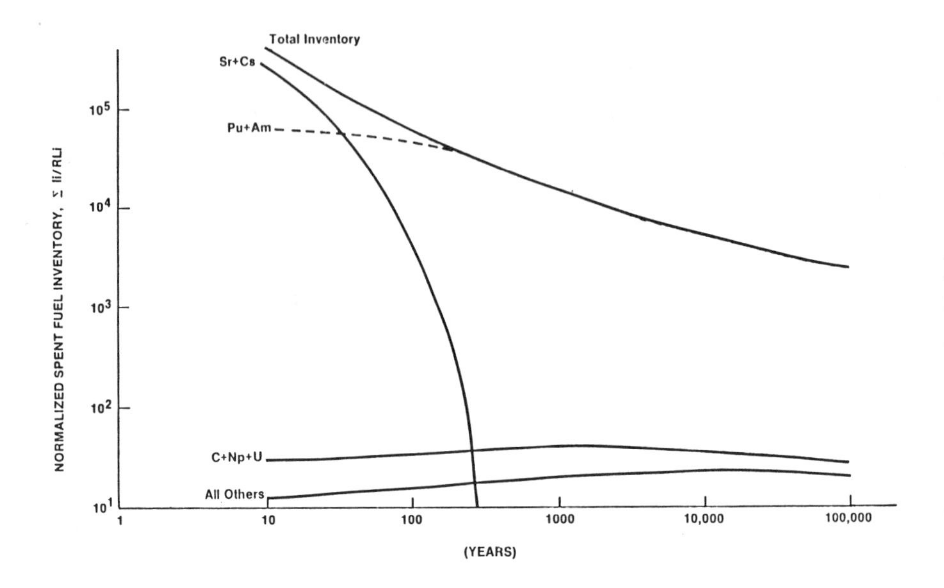

Figure 1. Spent fuel inventory normalized by release limits of 40 CFR 191.

M_{wp} where

$$M_{wp} = \Sigma Q_i/RL_i$$

and where Q_i is the cumulative release of the ith radionuclide and RL_i is the release limit of the containment requirements of 40 CFR 191.13. Formally the goal is that M_{wp} is 0.1 with medium confidence. The meaning of medium confidence in this case is that if the waste package release were a normally distributed variable, the probability that M_{wp} is less than 0.1 would be about 70 percent.

Meanwhile the goal for the natural barriers is based on the concept that, if the radionuclide travel time is sufficiently long, the fraction of the radionuclide inventory that could be released to the accessible environment in 10,000 years is small enough that the release limit would be met by a factor of 10 even neglecting any containment or isolation within the engineered barrier system. The requirements on the fraction that can be released can be estimated from Figure 1. This figure shows the value of M as a function of time, where

$$M = \Sigma I_i/RL_i$$

and I_i is the spent fuel inventory of the ith radionuclide in the repository. It can be seen that the inventory has a value of about 15,000 times the regulatory standards at 1000 years after closure, due largely (99.7%) to the plutonium and americium isotopes. The fraction of radionuclides that could be released before 10,000 years must therefore be limited to about one in 100,000 in order for the system performance goal to be met. This would be accomplished, for example, with a normally distributed radionuclide travel time with a mean value of 200,000 years and a standard deviation of 70,000 years. Such values are consistent with analyses for the Yucca Mountain site[2].

The primary natural barrier that will be relied upon to meet this goal is the Calico Hills unwelded tuff that underlies the repository horizon. Another unit that will be considered is the host rock, the Topopah Springs member of the Paintbrush tuff. This unit will be considered as a barrier held in reserve. Finally, transport in the saturated zone will also be evaluated and this same goal will be applied to the units in this zone, principally the saturated portions of the Calico Hills and the Topopah Springs units.

To meet the overall goal for each of these natural barriers, goals are set for the ground-water travel time and for the retardation factors. For the Calico Hills, the goal for the ground-water travel time, T, is 10,000 years with a probability of at least 98 percent. This goal is consistent with the preliminary analyses for this unit in the EA, and, as shown in Figure 1, this goal would be sufficient to delay all of the radionuclides with the exception of the plutonium and americium isotopes such that the fraction of the inventory of the former set of radionuclides would meet the desired goal since only 2 percent or less of the inventory could be released before 10,000 years. The goal for the retardation factor, R, is that R is greater than 5 with a high confidence for both Pu and Am. This goal, in concert with the ground-water travel time goal, is chosen to ensure that the fraction of Pu and Am that is released also meets the desired goal. That is, if the retardation factor were normally distributed with a probability of 98 percent of having a value greater than 5, than the fraction of the Pu and An inventory with a travel time less than 10,000 years would be on the order of 10^{-5}. Thus, the long ground-water travel time and the retardation of Pu and Am would ensure that the regulatory requirements would be met by an order of magnitude or more, even neglecting other favorable phenomena and the retardation of other radionuclides that may be in effect.

The goal set for the ground-water travel time in the Topopah Springs member is 5000 years with a high confidence. The preliminary analyses give a median (50% probability) ground-water travel time of about 5000 years. Thus, the present information only gives a low confidence that 5000 years would be met. The goal set for the retardation factor for C, U, Np, Pu, and Am is 10 with a high confidence. If the goals for both of these measures were met, the inventory for these elements released in 10,000 years would be more than an order of magnitude below the release limits on the basis of the same arguments given for the goals for the Calico Hills. If the ground-water travel time were at least 10,000 years with only a 50 percent probability, the remaining radionuclides would also meet the performance objective and it is assumed that the ground-water travel time goal specified is consistent with this need.

The preliminary analyses for the saturated zone give ground-water travel time estimates which are on the order of 100-1000 years. On the basis of arguments similar to those for the previous barriers, goals for the ground-water travel time of 100 years with a high confidence and for the retardation factor of 500 with a high confidence are set. The retardation factor goal must apply to virtually all elements in this case since the ground-water in this unit alone cannot be relied on to prevent release of any radionuclides before 10,000 years.

These considerations have so far applied to the ground-water transport of radionuclides. With regard to volatile radionuclides that might be subject to gaseous transport, the primary barrier to be relied upon is the engineered barrier system. In addition the overlying rock units and the shaft seals will be relied upon. The goal for the engineered barrier system is that the release of volatile radionuclides at the boundary of the waste package would be less than 10 percent of the release limits. The goal for the transport of the volatile radionuclides through the natural barriers is that the cumulative release in 10,000 years is less than 10 percent of the total inventory of these radionuclides. This goal would assure that the regulatory standard would be met by the volatiles. This goal could be accomplished by sufficiently long transport time or by substantial dispersion in the geology during transport. The shafts could provide a preferential pathway for gaseous transport of radionuclides. Therefore the goal for the shaft seals is that the transport properties are comparable to those of the adjacent rock.

Information Needs

The next step in the performance allocation process is to identify and set goals for the parameters needed for the evaluation of the performance measures in Table 1. This step must therefore be conducted for waste package, hydraulic, and retardation parameters and parameters for gaseous transport. For the purposes of illustrating this step, parameters are identified and goals are set associated with the ground-water travel time.

The parameters needed to evaluate the ground-water travel time in the specified lithostratigraphic units cannot be identified with certainty at the present time since the mechanisms for local ground-water movement at the site are not currently well understood. However, the simplified models consistent with existing information that have been used to support the preliminary ground-water travel time estimates in the EA are considered to be appropriate as a basis for the site characterization program in this respect.

The existing information discussed in Section 6.3.1 of the EA[1] suggests that two modes of ground-water movement should be considered in evaluating the ground-water travel time in the unsaturated portion of the Calico Hills unwelded tuff and the Topopah Springs member of the Paintbrush tuff: vertical matrix flow and fracture flow. The principal parameters needed to evaluate matrix flow in these units include the vertical percolation flux, the relative hydraulic conductivity of the matrix, the effective matrix porosity, and the thickness of these units. The EA used a simple, conservative model to evaluate the particle velocity, V_p in these units:

$$V_p = (q/n)(q/K_s)^{-E}$$

where q is the percolation flux, n is the matrix porosity, K_s is the saturated conductivity of the matrix and E is an empirical parameter that accounts for the effects of saturation and pore-size distribution on the particle velocity[3]. The parameters in this expression depend upon the characteristics of the water and the rock matrix.

For fracture flow in the unsaturated zone, the analysis in the EA assumed that the particle velocity could be estimated from the expression

$$V_p = (q-K_s)/n_e$$

where n_e is the effective porosity of the fractures in the unit and the other parameters have the same meaning as before. A value for the effective porosity of the fractures is variable and not well understood at present. A value of 0.0001 was used in the estimates in this case.

The ground-water travel time can be estimated once the particle velocity is known. For example, the EA estimated the travel time by numerically evaluating the travel time through the rock based upon the relationship:

$$T = \int dx/V_p(x).$$

The EA also presents a simple model for ground-water movement in the saturated zone below the water table in the vicinity of the site. In this case the flow is expected to be predominately in the fractures and estimates of the travel time were made assuming that the flow is essentially horizontal, darcian, and represented by equivalent porous media properties. In the vicinity of the site the water table extends into the Calico Hills nonwelded tuff and, for a portion of the travel path, into the dipping Topopah Springs member. Important parameters needed to evaluate the travel time in the saturated zone include the length of the travel path in the units, the hydraulic conductivity and gradient, and the effective porosity of the fracture system.

Consistent with the information available for the site summarized in the EA, the goals for the needed parameters in these units are given in Table 2. The goals chosen for the hydraulic parameters for the Topopah Springs and Calico Hills units are consistent with existing information and, if used in the models described, would be likely to provide general agreement with the goals for ground-water travel time.

The EA analyses do not explicitly consider special ground-water pathways that might be associated with distinct

Table 2. Goals for Hydraulic Parameters to Resolve Issue.

Parameter	Goal	Needed Confidence	Expected Values
Unsaturated Zone - Calico Hills Zeolitized Unit			
Saturated Hydraulic Conductivity	Ks < 8 mm/yr	Med	< 8
Percolation Flux	q < 0.5 mm/yr	Med	< .5
Effective Matrix Porosity	n > 0.2	Med	≈.25
Effective Fracture Porosity	ne > 0.0001	Med	--
Unit Thickness	t > 60 m	Med	0 - 133
Brooks-Corey Parameter	E > 7	Med	--
Unsaturated Zone - Topopah Springs Member			
Saturated Hydraulic Conductivity	Ks < 4 mm/yr	Med	< 4
Percolation Flux	q ≤ 0.5 mm/yr	Med	< .5
Effective Matrix Porosity	n ≥ 0.06	Med	≈.10
Effective Fracture Porosity	ne > 0.0001	Med	--
Unit Thickness	t > 30 m	Med	0 - 72
Brooks-Corey Parameter	E > 6	Med	--
Saturated Zone - Topopah Springs Member			
Saturated Hydraulic Conductivity	Ks < 300 mm/yr	Med	≤ 365
Horizontal Hydraulic Gradient	i < 0.1	Med	≈ .0001
Effective Fracture Porosity	ne > 0.0001	Med	> .001
Path Length	t > 1000 m	Med	≈ 1000
Saturated Zone - Calico Hills Unit			
Saturated Hydraulic Conductivity	Ks < 50 mm/yr	Med	< 59
Horizontal Hydraulic Gradient	i < 0.5	Med	< .001
Effective Fracture Porosity	ne > 0.0001	Med	> .001
Path Length	t > 4000 m	Med	≈ 4000
Seeps for Lateral Diversion in Topopah Springs and Calico Hills			
Volumetric Flow rate	Q 0.05 m^3/yr	Med	
Saturated Hydraulic Conductivity	Ks 10 m/yr	Med	--
Ghost Dance Fault			
Downward Flux	q 1 m/yr	Med	--

features at the site and which could affect the travel-time estimates. For example, the diversion of water at seeps in the unsaturated zone that might result in perching or down-dip flow at unit interfaces could affect the picture of the flow regime.

Likewise, downward flow through a conduit such as a major fault could also affect the travel time estimates. The hydraulic characteristics of such features should be tested. However, no simple model presently exists to evaluate specific effects on travel time and goals for the parameters are based on judgment.

The goals for these parameters have also been given in Table 2. A goal has been set for lateral diversion of the flow into a seep based upon the goal for flux of 0.5 mm/yr and an effective cross-sectional area of 100 m^2. For down-dip flow at the interface between the Topopah Spring and Calico Hills units a goal of 10 m/yr has been set for the saturated hydraulic conductivity of this interface. A goal has been set for the local downward flux in the Ghost Dance fault of one m/yr. It is believed that if these goals were met, the goal for the ground-water travel time would also be met at the site.

Also shown in Table 2 are the expected values for these parameters. Estimates of the existing confidence in the goals can be made based upon these expected values. Where the expected range of values appears to be easily met by the goal, the current confidence is assumed to be high. In cases where the current confidence is high relative to the needed confidence it is expected that extensive testing will not be necessary. In cases where little or no data exist or where the existing uncertainties are very large, current confidence may be low and the need for testing may be greater.

CONCLUSION

The approach to the resolution of issues confronting the program for characterization of sites for a mined geologic disposal facility has been presented. This performance allocation approach provides the framework for presenting the rationale for the testing and analysis to be conducted during characterization and provides the logical development of parameters to be measured from the performance objectives set for the repository system. An example is given to illustrate this approach and to show the development of performance goals, needed confidence levels, parameters, needed confidence levels in these parameters and, existing confidence levels.

REFERENCES

1. U.S. Department of Energy. Environmental Assessment For Yucca Mountain Site, Nevada Research and Development Area, Nevada (U.S. Department of Energy, Washington D.C., 1986).

2. Sinnock, Scott, Y.T. Lin, and J.P. Brannen. "Preliminary Bounds on the Expected Postclosure Performance of the Yucca Mountain Site, Southern Nevada", Sandia Report SAND84-1492 (Sandia National Laboratories, 1986).

3. Brooks, R.H. and A.T. Corey. "Properties of Porous Media Affecting Fluid Flow," Journal of Irrigation and Drainage Diversion, 92, No. IR2: 61-88 (1966).

MEASUREMENTS OF WATER INFILTRATION AND NUCLIDE RELEASES FROM LABORATORY MODELS OF ENGINEERED FACILITIES

Vern C. Rogers, Kirk K. Nielson, and D. Douglas Miller
Rogers & Associates Engineering Corporation,
Salt Lake City, Utah

ABSTRACT

Concrete and asphalt has been investigated for use in an effectively designed asphalt underground vault facility for low-level radioactive waste (LLW) disposal by performing bench scale tests comparing asphalt performance with concrete and earthen facilities.

The project involved the testing and evaluation of bench scale test models of various forms of concrete and asphalt vault LLW containment facilities for water infiltration and contaminant leaching characteristics. The facilities enclosed simulated waste material in the form of tracer salts, and were subjected to physical loading and differential loading tests to permit examination of their strength characteristics. Water flow and leach characteristics were determined before and after the loads were applied, to determine the effect of each load on each facility.

The tests revealed that the engineered barriers generally performed better than the earthen systems but that under some circumstances higher contaminant releases were observed from failed concrete systems compared to earthen systems.

Geotechnical and Geohydrological Aspects of Waste Management, D. J. A. van Zyl et al., Eds., © 1987 Lewis Publishers, Inc., Chelsea, Michigan—Printed in USA.

INTRODUCTION

Since the beginning of the nuclear age, disposal of Low Level Radioactive Waste (LLW) has been accomplished using Shallow Land Burial (SLB) techniques. However, within the past 15 years, radionuclide migration has been discovered in the vicinity of Maxey Flats, Kentucky, Sheffield, Illinois, and West Valley, New York commercial disposal facilities, and it has become apparent that these facilities have not performed as expected. These incidents, coupled with the great attention recently brought to bear on hazardous waste disposal failures, have resulted in a very low level of confidence from the general public and from state agencies in the capability of traditional SLB facilities to safely contain the waste materials. Among the findings of a 1984 NRC-sponsored workshop on alternative disposal concepts for LLW and SLB, was a feeling that the public placed greater confidence in disposal alternatives that make use of man-made engineered barriers [1]. It was further concluded that public perceptions of LLW disposal capabilities and the risks associated therewith are important, and that the added cost of technologies to increase public protection, and improve public perceptions is of secondary importance.

Additionally, a study was conducted by the Pennsylvania State University Institute for Research on Land and Water Resources to solicit the opinions of members of the general public in Pennsylvania concerning LLW disposal sites. When given four alternative methods of LLW disposal, the general public showed a preference for underground disposal with engineered barriers [2].

In the investigation of alternative technologies for LLW disposal, man-made engineered systems for LLW containment have been receiving a good deal of attention. Most notably, disposal technologies using concrete containment are being considered by most states and compacts, but there are concerns about the ability of a concrete structure to withstand an underground environment for an extended period of time.

Another candidate material, asphalt, may also be suitable for these purposes because of its many favorable characteristics, including better plasticity than other candidate materials, possession of annealing properties, better resistance to acid and chemical attack, and inherent water resistance. Century-old samples of asphalt have been recovered in excellent condition, attesting to the long term resistance of asphalt to the hardships of anaerobic burial.

A laboratory project, sponsored by the Department of Energy, was undertaken at Rogers and Associates Engineering Corporation to investigate the use of concrete and asphalt in an effectively designed underground vault facility for LLW disposal, and to perform limited bench scale tests comparing concrete and asphalt performance with earthen facilities.

Water infiltration and leaching tests were performed on bench scale models of various forms of concrete, asphalt and earthen vault LLW containment facilities. The facilities enclosed simulated waste material in the form of tracer salts, and were subjected to physical loading and differential loading tests to permit examination of their strength characteristics. Leach and water infiltration characteristics were determined before and after the loads were applied, to determine the effect of each load on each facility.

VAULT CONSTRUCTION

Three vaults of asphalt composition were involved in the testing. The first was composed of a cationic asphalt emulsion solids) admixture, with concrete sand and chopped fiberglass. The second was made by blending thermoplastic styrenebutadiene rubber (SBR) polymers with prime grades of asphalt, resulting in a "rubberized" asphalt with bulk properties similar to those of an elastomeric polymer. The third vault consisted of a concrete vault as substrate, coated with a thin layer of the SBR rubberized asphalt.

Two concrete vaults and two traditional earthen SLB facilities were also constructed. These model vaults contained the same wastes, and were subjected to the same environment and loads as the asphalt vaults.

The LLW test vaults were each constructed in a two-cell geometry, and were compacted into an earthen fill in metal test chambers as illustrated in Figure 1. Vault dimensions were determined by similitude modeling for mechanical performance, with a length scale of 67. The cells were approximately 12-cm wide, 9-cm deep, and 15-cm long (30-cm total), with 1.5-cm wall thicknesses. For the concrete vaults, a galvanized wire screen was cast into the walls to simulate reinforcing steel. The asphalt vaults were cast into forms identical to those used for the concrete cells. For the earthen facilities, the inner sections of the vault forms were used directly in the soil to compact a defining shape in the soil that corresponded in dimensions to the concrete and asphalt vaults.

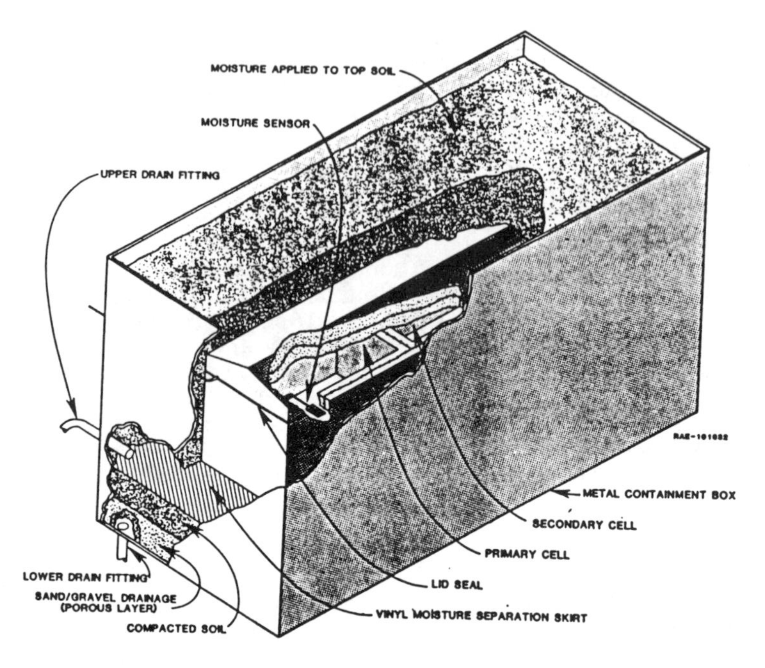

Figure 1. Diagram of model vault.

Each vault was housed in its own individual containment, a 25 cm x 25 cm x 46 cm sheet metal box. The boxes were drilled with holes to permit drainage of water from the bottom, and were fitted with a vinyl membrane to separate water seeping through the vault from water directed around the vault. A 1 cm layer of sand and gravel was first placed in the bottom of the boxes to enable water seeping through the vault to move quickly to the drain once it reached the bottom of the box. The base of each drain was surrounded by glass wool to prevent the soil from entering and clogging the drain.

Water seeping through the vault, and water directed around the vault were drained from two separate locations. For the water around the vault, a vinyl skirt was cemented to the sides of the vault and to the wall of the containment. This water entered the side drain, which was located in the same corner as the bottom drain. Thus all water which actually passed through the vaults was drained from the bottom of the containment, and all water which did not was drained from the side. The metal containment boxes were tilted to promote proper drainage.

The fill material was a local sandy soil, with a Proctor moisture of 8.5 percent and a Proctor density of 121.3 pounds per cubic foot. The soil was packed at optimum moisture, and at a density somewhat less than Proctor in order to avoid premature damage to the vaults.

Each cavity of the vaults was filled with simulated waste material, which consisted of tracer salts, ground to a powder, and mixed with soil. Two hundred grams of simulated waste was allotted to each cell and mixed with enough soil to fill the cell.

Salts were selected on the basis of their solubilities. One highly soluble salt was combined with a salt of much lower solubility for use in the same cell. In this manner, water going through the cell was easily detectable almost immediately, but all of the salts were not immediately flushed out of the system. Some of the salt of lower solubility would remain in the vault throughout the experiment; thereby permitting the detection of flow through the vault throughout the experiment. The simulated waste of the primary cell, that portion of the vault which bore the applied load, consisted of equal portions of Mercuric Iodide, Potassium Permanganate, and Calcium Tungstate. The secondary cell contained Thallous Bromide.

EXPERIMENT PLAN

The seven bench-scale test models of LLW vaults and the tests performed on them are listed in Table 1. The tests included moisture infiltration, leaching and mechanical stability.

The concrete and rubberized asphalt vaults were also compared on the basis of their resistance to sulfate attack. Naturally occurring sulfates are often present in soil or dissolved in groundwater near the vaults, and they can attack concrete. This problem is of particular concern in arid climates, but can occur anywhere [3]. The resistance of the vaults to sulfate attack was compared by using a 1 percent sodium sulfate solution instead of water as the leaching agent on one each of the concrete and asphalt vaults. This represents a severe to very severe exposure of the concrete to sulfates in the water [3].

The moisture infiltration of each test vault was monitored via electrical moisture sensors inside the vaults. Up to fourteen pore volumes of water or aqueous sodium sulfate solution were applied to the top

Table 1. Definition of the Seven Bench-scale LLW Containment Vaults.

Containment		Infiltration Test	Pressure Test
Earthen --	SLB1	Water	Yes
	SLB2	Water	No
Concrete --	CON1	1% Na_2SO_4	Yes
	CON2	Water	Yes
Asphalt --	Admix	Water	Yes
	Rubrasp	1% Na_2SO_4	Yes
	Aspcon	Water	Yes

of the earthen containments around the vaults, and water balance data (volume applied versus volume collected) were also obtained. Separate water collection ports were used for water penetrating through the vaults and water passing around the vaults.

The mechanical stability tests were initiated after the fourth pore volume of water had been applied and retrieved from each test system. These utilized a hydraulic press to apply increasing pressures in successive tests of 30 to 1000 pounds per square inch on a 40 square-inch pressure plate located over one of the two chambers in the test vault. These corresponded to total loadings of 1200 pounds to 40,000 pounds (540 kg to 18,000 kg). After each loading test, an additional water infiltration test of 1-2 pore volumes was conducted to examine possible structural damage from the loading tests.

Waste contaminant and leaching was estimated by collection of the water from the infiltration tests and analyzing for chemical elements that were spiked into the simulated wastes in the model vaults. The primary cell of each vault (to which the pressure loading was applied) contained approximately 1 g of $KMnO_4$, 1 g of HgI, and 1.6 g of $CaWO_4$. The secondary cell contained approximately 1 g of TlBr. Leachate solutions were analyzed by x-ray fluorescence for K, Mn, Hg, I, Ca, W, Tl, and Br.

RESULTS OF THE LOADING TESTS

After the loading and leaching tests were completed, the soil around the vaults was removed and the vaults were disassembled for visual observations of their physical condition.

Concentrations in the leachate were very similar in SLB 1 and SLB 2. At higher pressures, however, the salt concentrations in leachate from SLB 1 increased over those from SLB 2.

There was definite failure in all of the concrete vaults, characterized by sudden inability to support the pressure of the load, and audible cracking noises. Asphalt vaults generally showed an ability to deform under the loads, with no definite point of failure. The asphalt-concrete (aspcon) vault behaved structurally as a concrete vault, but its leach properties were quite different, as shall be discussed later.

Concrete I, which was leached with a sulfate solution, failed at approximately 80 psi. The structural failure of the vault was quite complete, including cracks in the corners extending to well below the level of the skirt, thereby permitting the escape of salts to the bottom drain. Cracks also crossed the bottom of the primary cell. The failure of the roof included a wide gap which extended to a corner of the secondary cell, which permitted water to enter the secondary cell. This cell was filled nearly to the top with water.

Concrete II failed at approximately 150 psi. Comparison with the failure of Concrete I reveals the sulfate solution weakened the concrete. Once again, failure was quite complete, with wide cracks in the corners extending to the bottom of the vault.

The admix vault deformed quite well under the loads. There was no definite, observable point of failure. The admix appeared structurally capable of supporting the loads if necessary. However, disassembly of the vault after completion of the tests revealed gaps in the corners of both cells, probably caused by the heavier loads. These gaps provided drainage to both drains. Also, the underside of the bottom was cracked at the location of the central wall. Soil in both cells was wet, but there was no standing water. The rubberized asphalt (rubrasp) vault showed no structural failure, but continuously deformed under the loads, and maintained its structural integrity.

The simulated waste of the primary cell was dry, while that of the secondary cell was wet, indicating that the cap was probably not perfectly sealed to the vault. No cracks or holes were visible in the walls, floor, or roof of the vault.

The asphalt-concrete (aspcon) vault behaved in a similar manner as Concrete II, sustaining loads up to approximately 150 psi, where it failed suddenly. Removal of the vault from the model showed that failure of the concrete had been quite catastrophic, including a crack of 6-10 mm width extending the entire length of the primary cell, through the central wall, and well into the secondary cell. Visual inspection of the vault, however, revealed no gaps in the asphalt coating despite catastrophic failure of the substrate. The simulated waste material inside the vaults was completely dry.

The aspcon vault failed at about the same point as Concrete II. However, the rubrasp membrane was conformed to the new deformed shape of the concrete substrate, preserving its water-tight integrity.

WATER INFILTRATION AND LEACH TESTS

Typical salt concentrations in the leachate from representative test chambers are given in Figures 2 through 4. The solid circles in these figures represent concentrations that were measured above the detection limit. The arrows indicate that the element concentration was below the detection limit, and the open circles indicate no leachate was collected.

As observed in Figure 2, the shallow land burial models performed as expected, with leachate

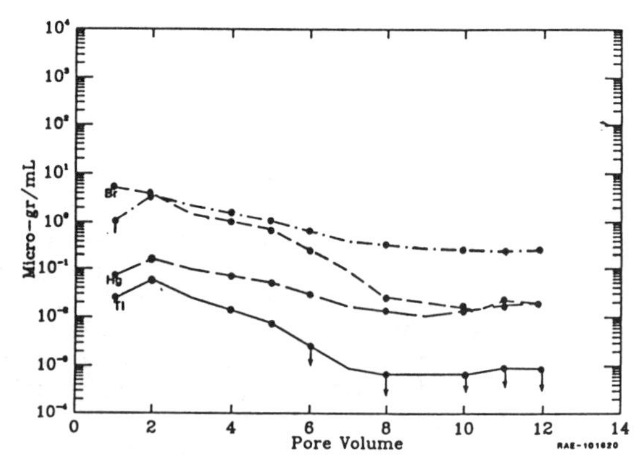

Figure 2. SLB1 leachate history.

concentrations being at their highest value, initially, and gradually diminishing as the amount of salts in the trenches decreased.

Salt concentrations in the leachate from the concrete vaults remained quite low until vault failure from the high pressure load tests. The concentrations in the side drain increased suddenly with each loading, as though the lightest loads caused some compromise of the integrity of the top and sides of the vault. For the bottom drain data, given in Figure 3, the salts of the primary cell (Hg and I) remain low in concentration until they begin to increase after PV8, indicating that there may have been some unnoticed failure under the 120 psi load which immediately followed PV8. Meanwhile, salts of the secondary cell remain low in concentration until after the catastrophic failure which followed PV10. After this, the infiltration of the secondary cell, and the leaching of the salts of the secondary cell became evident. From the initial concentration, the salts of the primary cell generally declined in concentration until heavier loads were placed upon the cell. Finally, the concentrations of these salts increased dramatically upon failure of the vault. The salts of the secondary vault fluctuated in concentration over a range of approximately one order of magnitude, but remained relatively consistent throughout.

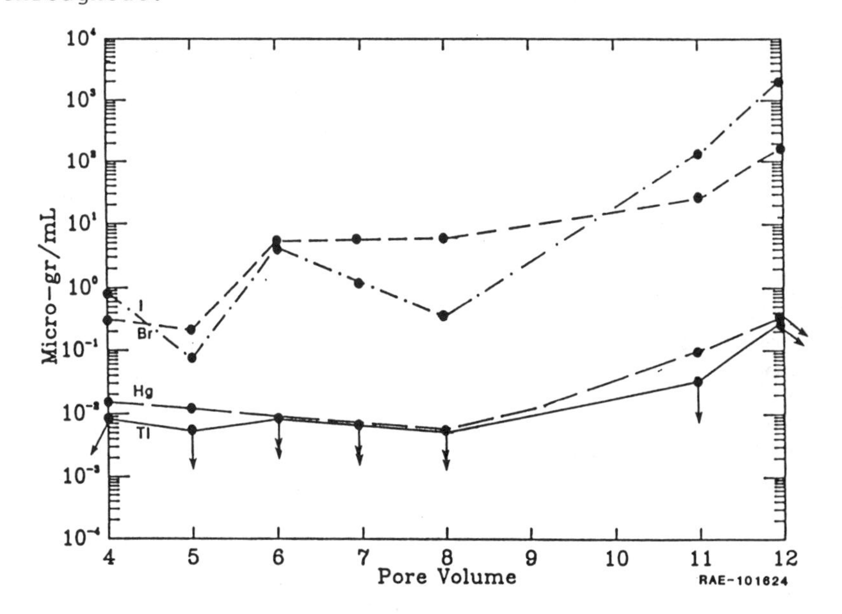

Figure 3. CON2 bottom drain leachate history.

Just as with the rubrasp vault, the performance of the aspcon vault doesn't degrade significantly with loading. Concentrations in both side and bottom drains are consistently in the 10^{-3} micro-gm/ml range, with occasional increases to the 10^{-2} range. Even upon failure of the vault the concentrations of the most soluble elements only reached the 10^{-2} range in the side drain. The bottom drain did not even see any leachate for most of the experiment.

The total integrated salt releases are summarized for all of the test models in Table 2. These indicate that major fractions of the total salt inventories were released from the SLB1, SLB2, CON1, CON2, and Admix vaults. The improved performance of the Rubrasp and Aspcon vaults is also shown by these data. In general, the Aspcon vault exhibited the best overall performance in the leaching tests.

Table 2. Total Integrated Salt Releases From Side and Bottom Drains of the Seven Test Vaults (grams).

	Hg	I	W	Tl	Br
Bottom Drains					
SLB1	0.0034	0.062	0.00069	0.00079	0.073
SLB2	.0057	.156	.00072	.0011	.126
CON1	.0011	.10	.0029	.0011	.0037
CON2	.00013	.42	.0010	.00007	.042
Admix	.0021	.13	.0014	.00032	.069
Rubrasp	.00039	.0012	.0017	.00021	.0013
Aspcon	.00018	.0017	.00049	.00009	.00057
Side Drains					
CON1	.00058	.057	.0017	.00034	.013
CON2	.0012	.037	.00070	.00013	.0042
Admix	.00011	.0015	.00032	.00004	.0069
Rubrasp	.00074	.0027	.0031	.00047	.0015
Aspcon	.00005	.00028	.00024	.00004	.00070

Water had collected in the secondary cell, but was restricted to a depth of not more than 5 mm, indicating that either there was ample opportunity for the water to drain, or the amount of water entering the secondary cell was limited. Water was not permitted to collect in the primary cell, but drained easily from the substantial gaps created in the cell corners by the loading of the cell.

Moisture detectors in the Concrete vault indicated that there was moisture present in both cells throughout the experiment.

The rubrasp vault's performance was very consistent throughout the experiment, with concentrations mainly in the 10^{-2} to 10^{0} micro-gm/ml range. Bromine and Iodine increased briefly up to 10^{-1} range, but returned immediately to 10^{-2}. For the bottom drain, given in Figure 4, the concentrations of elements from both cells began rising after PV5. However, the concentrations then either stabilized (as with Bromine and Iodine), or began to decrease (as with Mercury and Thallium). The later, heavier loads showed no ability to compromise the integrity of this vault.

Disassembly of the vault revealed a dry primary and a wet secondary cell. Water was not standing in the secondary cell, but the soil was moist.

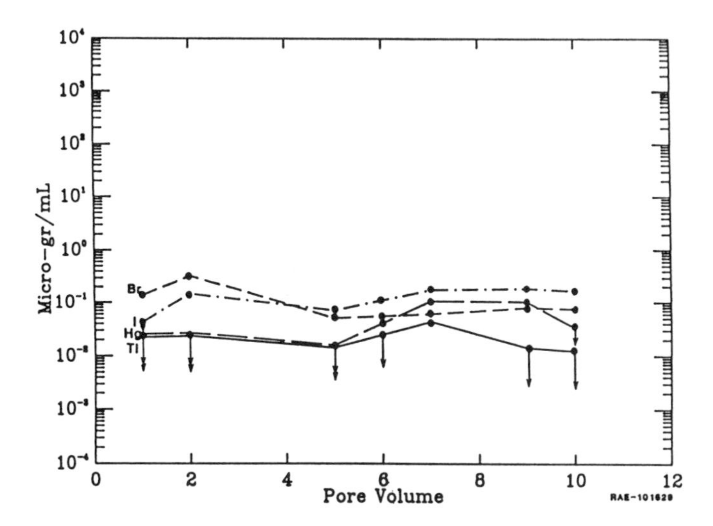

Figure 4. RUBRASP bottom drain leachate history.

REFERENCES

1. "Proceedings of the State Workshop on Shallow Land Burial and Alternative Disposal Concepts," Nuclear Regulatory Commission report, NUREG/CP-0055 (1984).

2. Bord, R.J., "Opinions of Pennsylvanians on Policy Issues Related to Low Level Radioactive Waste Disposal," The Pennsylvania State University Institute for Research on Land and Water Resources (1986).

3. "ACI Manual of Concrete Practice, Part 1," American Concrete Institute, Detroit (1985).

EXCURSION CONTROL AT
IN SITU URANIUM MINES

William P. Staub,
Oak Ridge National Laboratory, Oak Ridge, Tennessee

INTRODUCTION

Intensive research and development of in situ uranium mining took place in the United States during the 1970's. By the end of the decade, nearly 10% of all uranium production came from in situ mines. Recent poor market conditions, however, forced the closure of all domestic in situ uranium mines.

When market conditions improve, domestic in situ mining is expected to make a modest recovery. Successful licensing of future in situ mines will depend to a large extent on the ability to contain the leach solutions (lixiviant) within the ore zone. Uncontrolled movement of lixiviant beyond the ore zone is called an excursion.

Early detection of an excursion is a necessary prelude to the implementation of timely corrective action to return escaping fluid to the well-field. Without timely corrective action, a large quantity of valuable production fluid may be irretrievably lost and serious contamination of adjacent groundwater resources may result.

Two types of excursion are recognized: horizontal and vertical. In horizontal excursions lixiviant remains strata bound but migrates laterally away from the production well-field. In vertical excursions lixiviant escapes into aquifers above or below the ore bearing strata.

In situ mining may not be permitted when monitor wells outside the ore zone are on excursion status. Monitor wells are classified as being on excursion status whenever selected chemical constituents (excursion indicators) exceed given concentrations, referred to as upper control limits (UCL's).

This paper summarizes excursions based on case histories of 8 in situ uranium mines (7 in Wyoming and 1 in Texas). These case histories were compiled from data provided by the U.S. Nuclear Regulatory Commission, the Wyoming Department of Environmental

Geotechnical and Geohydrological Aspects of Waste Management, D. J. A. van Zyl et al., Eds., © 1987 Lewis Publishers, Inc., Chelsea, Michigan – Printed in USA.

Quality, and the Texas Department of Water Resources. Most of these data were provided to the above agencies by mining companies in response to regulatory requirements pertaining to licensing actions. Case histories and detailed excursion analyses are presented in Staub et al. [1]

HORIZONTAL EXCURSIONS

The only serious horizontal excursions described by Staub et al. [1] occurred during the early history of in situ uranium mining. Several intense and long-term excursions took place during experimental pilot tests in the early- to mid-1970's. These excursions were attributed to numerous breakdowns in pumping equipment and failure to adjust injection-production rates in response to these breakdowns. The total volume of fluid injected during these pilot tests exceeded the amount produced, a practice which is now recognized by the industry as a fundamental cause of horizontal excursions.

The in situ mining industry has had a good record of controlling horizontal excursions since the late 1970's. Most recent horizontal excursions have been brought under control within 2 to 5 months of their discovery. The incidence of horizontal excursions is reduced by controlling the field-wide production rate at a few percent greater than the injection rate (known as "bleeding" the ore zone aquifer) and storing the excess fluid in a secure surface impoundment for eventual evaporation. Although bleeding the aquifer prevents field-wide excursions, local excursions occasionally occur. They are controlled by either local or field-wide manipulations of injection and production rates ranging from slightly reduced injection to complete injection well shut-down while continuing to produce from the ore zone aquifer. The range of options available depends on the size of the impoundment for storage of excess production fluid.

VERTICAL EXCURSIONS

To date, the in situ mining industry has had difficulty in controlling vertical excursions. Serious vertical excursions persisted at several sites through the end of 1981 when commercial scale in situ mining was suspended in response to the declining uranium market. A number of shallow aquifer monitor wells have remained on excursion status for several years.

Many vertical excursions are attributable to broken casings in injection or production wells or improperly abandoned exploration wells. State-of-the-art well completion procedures have virtually eliminated injection and production well failures. Old and abandoned exploration wells remain as major pathways of excursion because they are difficult to locate.

Some vertical excursions may also be responses to leaky aquifer conditions [2]. Too little is known at this time to assess the significance of leaky aquifers relative to that of abandoned wells in causing vertical excursions.

Vertical excursions may be controlled by repairing ruptured casings in injection and production wells or, alternatively, by abandoning and sealing them off with cement. Improperly sealed or unsealed exploration holes are located and also sealed off with cement. Once sealed, these wells cannot be used to recover escaped lixiviant. Restoration of the shallow aquifer may be required, depending on the extent of the excursion. Completely new well-fields would need to be developed for restoration. At present no techniques are available for controlling excursions attributable to leaky aquifer systems.

EVALUATING EXCURSION POTENTIAL

Although current aquifer testing procedures are often useful in appraising the potential for horizontal excursions, they have had less success in determining the potential for vertical excursions. The short duration of these tests and the small number of wells involved reduces the possibility of measuring observable responses in shallow aquifer monitor wells. Furthermore, such tests reflect only localized rather than field-wide conditions.

To overcome the disadvantages of short-term aquifer tests a mining unit could be certified by commencing operations but injecting a chemically stable and unretarded tracer instead of lixiviant. Production fluid could be held in surge tanks or an evaporation pond until the test is completed. Shallow monitor wells would be observed for changes in water level and possibly the appearance of tracers. An appropriate duration for certification tests is uncertain at this time. Each time a vertical excursion occurs the mining unit would require recertification.

EXCURSION DETECTION

Both horizontal and vertical excursions can be more effectively controlled through timely detection. Effective excursion monitoring requires detailed groundwater chemical characterization to define natural variations in water quality. According to Deutsch et al. [3] median and variance in baseline water quality can be used to establish upper control limits (UCL's) for given ions based on their low probabilities of natural exceedance.

Proper selection of excursion indicators is critical. Table 1 summarizes the generalized suitability of various excursion indicators for monitor well observations, sampling and analysis. Some indicators are time dependent. Trace elements, TDS, and Cl^- are not concentrated until the lixiviant has circulated several times through the mining unit and ion exchange towers. Other indicators are lixiviant dependent (pH, alkalinity, and HCO_3^-). Any chemical indicator may be unsuitable when its baseline concentration is high and/ or has a broad range in the natural environment or is chemically unstable [3]. Chemically unstable constituents in solution are readily adsorbed on clay minerals or may precipitate if they are sensitive to changes in pH or oxidizing - reducing conditions. Water level may not be a reliable indicator where it

Table 1. Generalized suitability of excursion indicators.

Indicators	Potential Unique Interfering Chemical Reactions or Conditions	Time Dependency			
		Initial Lixiviant		Chemically Stabilized Lixiviant	
		Acid	Alkaline	Acid	Alkaline
TDS	None			X	X
Alkalinity	Subject to buffering by host rock and natural groundwater		X		X
Cl^-	None			X	X
HCO_3^-	Subject to buffering by host rock and natural groundwater		X	X	X
pH	Subject to buffering by host rock and natural groundwater	X		X	X
$SO_4^=$	Subject to chemical precipitation as gypsum	X		X	X
Trace Elements	Subject to chemical precipitation and adsorption on clay minerals			X	X
Water Level	Subject to external influences such as mine dewatering and irrigation	X	X	X	X

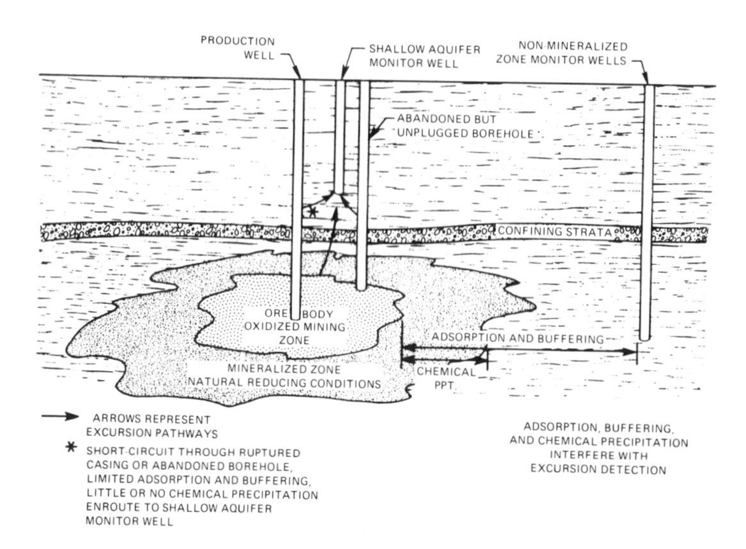

Figure 1. Comparison of pathways for horizontal and vertical excursions.

is subject to external influences such as mine dewatering, irrigation, and seasonal fluctuation.

Chemical instability is a significant concern when monitoring for horizontal excursions. As Figure 1 illustrates, vertical excursions often follow short-circuited pathways through oxidized zones. Hence, most ionic species tend to remain in solution so long as the monitoring well is screened in an oxidizing zone. Conversely, horizontal excursions often follow an extended pathway which includes a mineralized zone under natural reducing conditions. Thus, lixiviant escaping laterally along ore-bearing strata is subject to greater buffering, adsorption, and chemical precipitation than is lixiviant escaping through broken casings and abandoned but open exploration holes. Therefore, great care must be exercised in selecting indicators for monitoring horizontal excursions. In view of the above discussion, it is somewhat paradoxical that the in situ mining industry has a better track record in dealing with horizontal, rather than vertical, excursions.

There are two reasons why the industry's history of monitoring for vertical excursions has been less effective than one might expect, based on the above discussion. First, some shallow monitor wells were screened at locations where reducing conditions existed. Second, UCL's for trace element indicators were often set unrealistically low.

Monitoring within an aquifer where reducing conditions exist is problematical. Chemical indicators oxidized and solubilized by lixiviant in the ore zone may be reduced and precipitated along the pathway toward the monitor well. Natural concentrations of trace elements are often high where reducing conditions exist.

Introduction of oxygen during well construction and sampling may solubilize them.

Unrealistically low UCL's for trace elements often led to their abandonment as excursion indicators. The UCL's were typically set at 10 to 20% above baseline concentrations for both common ions and trace elements. In a case where the baseline concentration of uranium was 10 ugl, its UCL would be only 11 or 12 ugl (1 or 2 ugl above background). Because of unrealistically low trace element UCL's, monitor wells screened in a reducing zone often remained on excursion status for extended periods of time. Eventually, it was ascertained that these were not really excursions.

CONCLUSIONS

Based on the case histories in this study it is believed that the in situ uranium mining industry has demonstrated the ability to reduce and control horizontal excursions. However, the ability to reduce and control vertical excursions has yet to be demonstrated.

Detection of vertical excursions can be improved by screening monitor wells in oxidizing zones and using trace elements (for example uranium, vanadium, arsenic, selenium, and molybdenum) as excursion indicators. To provide reliable indications of an excursion, the UCL's should be set at more practical levels for trace elements, perhaps an order of magnitude above baseline concentrations. Trace elements can be very useful excursion indicators under appropriate conditions and guidelines.

Control of vertical excursions could be improved by mining unit-wide certification using operational mode tests without lixiviant before mining is allowed to commence. Each time a mining unit is placed on excursion status, recertification would be required.

ACKNOWLEDGMENTS

This research is sponsored by the Nuclear Regulatory Commission, Division of Waste Management, Office of Nuclear Material Safety and Safeguards under contract with Martin Marietta Energy Systems, Inc.

Special recognition is given to R. E. Williams, F. Anastasi, J. Osiensky and D. Rogness of the University of Idaho for compiling the case histories. The author is solely responsible for the excursion analysis based on these case histories.

REFERENCES

1. Staub, W. P., N. E. Hinkle, R. E. Williams, F. Anastasi, J. Osiensky, and D. Rogness. "An Analysis of Excursions at Selected In Situ Uranium Mines in Wyoming and Texas," USNRC Report ORNL/TM-9956 (1986).

2. Lee, D. W. and J. M. Bownds. "Hydrodynamics of Partially Penetrating Wells in a Leaky Aquifer System," USNRC Technical Letter Report ORNL/NRC LTR-86/14 (1986).

3. Deutsch, W. J., W. J. Martin, L. E. Eary, and R. J. Serne. "Methods of Minimizing Ground-Water Contamination from In Situ Leach Uranium Mining," USNRC Report PNL-5319 (1985).

QUALITY OF LIFE AND COMMUNITY SATISFACTION IN PROXIMITY TO HAZARDOUS WASTE

R. Gary Williams
Hazards Assessment Laboratory, Colorado State University

S. Jay Olshansky
National Opinion Research Center, University of Chicago

ABSTRACT

The NIMBY Syndrome (Not In My Back Yard) characterizes the social and political problems associated with siting hazardous waste facilities. Given a rational choice, everyone would prefer that hazardous wastes be located somewhere other than in their own backyard. While there has not been enough research that addresses the social and political effects of having a hazardous waste site located near communities, there have been qualitative case studies, anecdotal evidence, and environmental disasters such as Times Beach and Love Canal that would lead one to believe that hazardous waste sites are disruptive to communities. Media coverage of hazardous waste sites would lead one to believe that the majority of people in proximity to such sites are distraught, economic development in the area is negatively effected, property values decline, and in general, satisfaction with one's community suffers and quality of life decreases. Yet, social science research on this topic is essentially nonexistent. In fact, to date there is no published research that puts hazardous waste into the larger theoretical context of community satisfaction and quality of life.

Past research indicates that there are a number of dimensions to community satisfaction that include, among others, economic opportunity, availability of public facilities and services, integration into community, community cohesion, crime, etc. By placing issues associated with hazardous waste in a broader context, the relative importance of the hazardous waste on community satisfaction may be evaluated.

In this research, survey data were collected in an urban area in Northern Illinois that has a hazardous waste site

Geotechnical and Geohydrological Aspects of Waste Management, D. J. A. van Zyl et al., Eds.,

located in the center of the town. The study was designed to determine, in part, the unique effects of radioactive hazardous waste on overall levels of community satisfaction.

INTRODUCTION

In 1931, the Lindsay Light and Chemical Company moved its chemical operations for producing gas lamp mantels to a factory site in West Chicago, Illinois, about 30 miles west of Chicago. They began producing thorium at the West Chicago site in 1936, and subsequently provided thorium ore to the Manhattan Engineer District in the 1940s. In 1954, Lindsay Light and Chemical expanded their plant for the purpose of producing thorium nitrate, and they continued production until 1963. American Potash and Chemical Corporation purchased Lindsay Light and Chemical in 1958, and through a merger, the Kerr-McGee Corporation acquired the West Chicago facility in 1967. Due to adverse economic conditions, the Kerr-McGee corporation terminated the operation of the West Chicago facility in December of 1973.

The disposal of radioactive thorium wastes at the West Chicago site began in the early 1930s. While most of the wastes were buried on site, there have been occasions over the past 50 years when some of the contaminated soils have been used for landfill by local residents. In July of 1976, the U.S. Environmental Protection Agency found relatively high levels of radiation at the site and at nearby properties, and subsequently blocked off and closed sections of a nearby park. Through media coverage of these events, the public became aware of the existence of the radioactive wastes at the site, and discovered for the first time that some of their properties may contain soils that were previously contaminated by wastes from the facility.

In 1984, Kerr-McGee initiated a survey program with the city of West Chicago to determine the level of contamination in nearby properties, including the Park and the sanitary treatment plant. Surveys were subsequently completed on 2,726 properties, and close to 35 thousand cubic yards of contaminated soils have been excavated from 115 properties. These contaminated soils are now stored at the West Chicago site. The site is within the city limits of West Chicago and surrounded on three sides by residential areas. There is also a school approximately one block away from the site. The site is currently fenced and the industrial portion of the site, the part nearest the residential areas, has been reduced to a large rubble pile.

The West Chicago low-level radioactive waste site is particularly interesting from the perspective of community satisfaction. The site has been located in the middle of a densely populated urban area for over 50 years, and only within the last 10 years has the public been aware of the nature of the contamination. Many of the local residents have personal ties to the site in the form of previous employment of some family member, and the facility and its various operations have contributed to the local economy in the past. This West Chicago site is considered a natural laboratory of sorts, one in which community attitudes may be evaluated in the context of proximity to an area that is known to contain hazardous waste, and which is in the process of being evaluated for the possible removal of those wastes.

The literature on community satisfaction suggests that community attributes, interpreted by individuals, influences ones evaluation of community. It was expected that the waste site would be a salient community attribute which would negatively effect individual perception of community satisfaction. For the purpose of this study five hypotheses concerning the effects of waste on community satisfaction were tested. These included:

Hypothesis 1: People living in proximity to hazardous waste will be dissatisfied with their community.

Hypothesis 2: People dissatisfied with their community will be dissatisfied because of the hazardous waste.

Hypothesis 3: Perception of the health risks of hazardous waste and perception of the environmental effects of hazardous waste will produce stress, which will lead to lower satisfaction with community.

Hypothesis 4: Perception of low or reduced property values will lead to lower satisfaction with community.

Hypothesis 5: Lower community satisfaction will lead to consideration of migration from the community.

The last three hypothesized relationships are summarized in the path model in the results section of this paper.

METHODS

A five-page questionnaire was developed using standard indicators for stress [1], hypochondriasis [2], and community satisfaction [3]. Baseline data on general demographic characteristics of the population were also collected. A few initial interviews were conducted by telephone to pretest the questionnaire and the respondents' willingness to answer five pages of questions over the telephone. It was decided for better cooperation and sampling, the interviews would be conducted in person. The pretested sample of telephone interviews were included in the sample.

The sampling frame for the personal interviewing was defined as a three-block radius around the perimeter of the site (approximately one-half mile radius). Seventy people were interviewed over a period of five days in July 1986. Local residents were cooperative and very interested in the research, with only two people refusing to be interviewed. Interviews lasted 15 minutes on the average.

Because there is a large Mexican-American sector in the community, two separate questionnaires were developed. The first was the five-page instrument (in English) discussed above, the second was a one-page questionnaire written in Spanish for interviews with respondents that spoke no English. The second questionnaire represents a condensed version of the English language questionnaire. Four Spanish-only interviews were conducted and are included in the data set.

The data were analyzed using SPSS PC+ [4]. Three statistical procedures were used in analyzing the data: crosstabulations; correlations; and multiple regression. The primary technique was regression analysis, performed with backward deletion of the variables in the equation. This approach analyzes the effects of all the predictor variables together in accounting for variance in the dependent variable, and then deletes predictor variables one by one based on the partial correlation coefficients and associated F values. The least important variables in the equations are deleted first in this process. The resulting equation represents the variables that significantly contribute to explaining the variance in the dependent variable.

Three dependent variables were used in regression equations. Of the three dependent variables, community satisfaction and quality of life were measured on five-point Likert scales. These two variables were assumed to be continuous for purposes of analysis. The third dependent variable, desire to migrate, was a dichotomous variable. Independent variables were continuous parameters, some of which were recoded into new categories.

RESULTS

Tables 1 and 2 summarize general perception of community satisfaction and overall quality of life of residents living within three blocks of a radioactive waste site. As Table 1 indicates, 71 percent of the respondents were either very satisfied (30.4%) or somewhat satisfied (40.6%) with their community. Less than 20 percent of the respondents were dissatisfied with their community, and of those who were dissatisfied, less than 10 percent listed the radioactive waste site as the reason. Since the sample population lives within three blocks of the waste site, we hypothesized that levels of dissatisfaction with community would be higher than was demonstrated here. Specifically, we expected that dissatisfaction would be focused on the waste site. Neither of these hypotheses was supported by the data.

Table 1 Community Satisfaction In Proximity To Radioactive Waste

	FREQUENCY	PERCENT	CUM. PERCENT
Very Satisfied	21	30.4	30.4
Somewhat Satisfied	28	40.6	71.0
Neutral	7	10.1	81.1
Somewhat Dissatisfied	11	15.9	97.0
Very Dissatisfied	2	2.9	99.9

(How satisfied are you overall with your community?)

What do you like least about your community?

(fill in the blank question)

	FREQUENCY	PERCENT
Hazardous Waste	18	26.1
Services	16	23.2
Social	12	17.4
Other	12	17.4
Politics	5	7.2
No Response	7	8.7

When asked what residents liked least about their community, 26 percent responded the radioactive waste site, followed closely by 23 percent indicating problems with various community services. Other areas least liked included social aspects of the community (such as changing ethnic composition) and politics. It is interesting to note, however, that of those listing the radioactive waste site as the least liked aspect of their community, approximately 67 percent were either somewhat satisfied or very satisfied overall with their community.

In response to a general comparative question about quality of life in West Chicago, almost 29 percent thought life was better or much better than other cities and towns in the U.S. (see Table 2). Over 62 percent of the respondents thought the quality of life was about the same in West Chicago as in other cities and towns in the U.S., while 9 percent of the sample thought the quality of life was worse.

Variation in the quality of life indicator could not be accounted for by any of the predictor variables discussed in the hypotheses. We assumed that the lack of variation in the indicator of quality of life is the primary reason that no statistically significant relationships emerged.

A number of demographic variables were used to analyze the potential effects of radioactive waste on perceptions of community satisfaction. These variables are included in the correlation matrix below (Table 3). There is only one statistically significant correlation with respect to overall satisfaction. Specifically, the less satisfied one was with community, the more likely that person was to have considered moving from the area.

Regression analysis was used to test the effects of length of residence, education, and sex on community satisfaction. Of these variables, only length of residence showed a statistically significant relationship to community satisfaction ($p < .05$). Length of residence accounted for 6.3 percent of the variance in community satisfaction.

The relationship between stress and community satisfaction was also tested using regression analysis. No statistically significant relationship existed between stress and community satisfaction when analyzed in a simple regression equation.

Table 2 Overall Quality of Life in Proximity to Radioactive Waste

	FREQUENCY	PERCENT	CUM PERCENT
Much Better	4	6.1	6.1
Better	15	22.7	28.8
About the Same	41	62.1	90.9
Worse	6	9.1	100.0
Much Worse	0	0	

(Overall, how do you rate the quality of life in your community as compared with what you think is the average situation for other cities and towns in the United States?).

Table 3 Correlation Matrix of Variables in the Regression Equations

	RESLNG	SEX	EDUC	HTHRSK	NUCWST	STRESS	PROP	SATIS	MIGR
RESLNG	1.00								
SEX .02	1.00								
EDUC	-.18	.12	1.00						
HTHRSK	.08	.03	.01	1.00					
NUCWST	.15	.19	.03	.57**	1.00				
STRESS	.14	.02	-.27	.25	.27	1.00			
PROP	.19	-.24	-.01	-.09	.04	-.11	1.00		
SATIS	.26	-.13	.15	-.09	-.02	-.15	.12	1.00	
MIGR	.23	-.02	-.01	.23	.17	.19	.05	-.48**	1.00

*1 tailed significance at .01 probability level

**1 tailed significance at .001 probability level

It was hypothesized that community satisfaction would be an important predictor of the desire to migrate. In a simple regression equation, community satisfaction accounted for 20 percent of the variance of the desire to migrate variable ($p < 0.05$).

The conceptual model of community satisfaction and desire to migrate is presented below. This model represents how perceived health risk associated with radioactive waste and perceived environmental effects of radioactive waste, acting through stress, could affect one's evaluation of community satisfaction. In turn, one's evaluation of community satisfaction could affect one's desire to migrate. Other variables expected to be important in determining community satisfaction are also shown in the model and include demographic characteristics and perceived property values. This model was tested systematically. The findings were not consistent with the hypothesized relationships. Perceived health risk of radioactive waste and perceived environmental effects of radioactive waste did not significantly contribute to stress, nor did stress significantly contribute to community satisfaction. In addition, perceived property value change did not significantly affect one's evaluation of community satisfaction. Consistent with the literature on community satisfaction, length of residence did contribute to explaining the variance associated with that variable although the explained variance was small (6.3 percent). Community satisfaction was a significant predictor of desire to migrate

Conceptual Model of Community Satisfaction and Migration

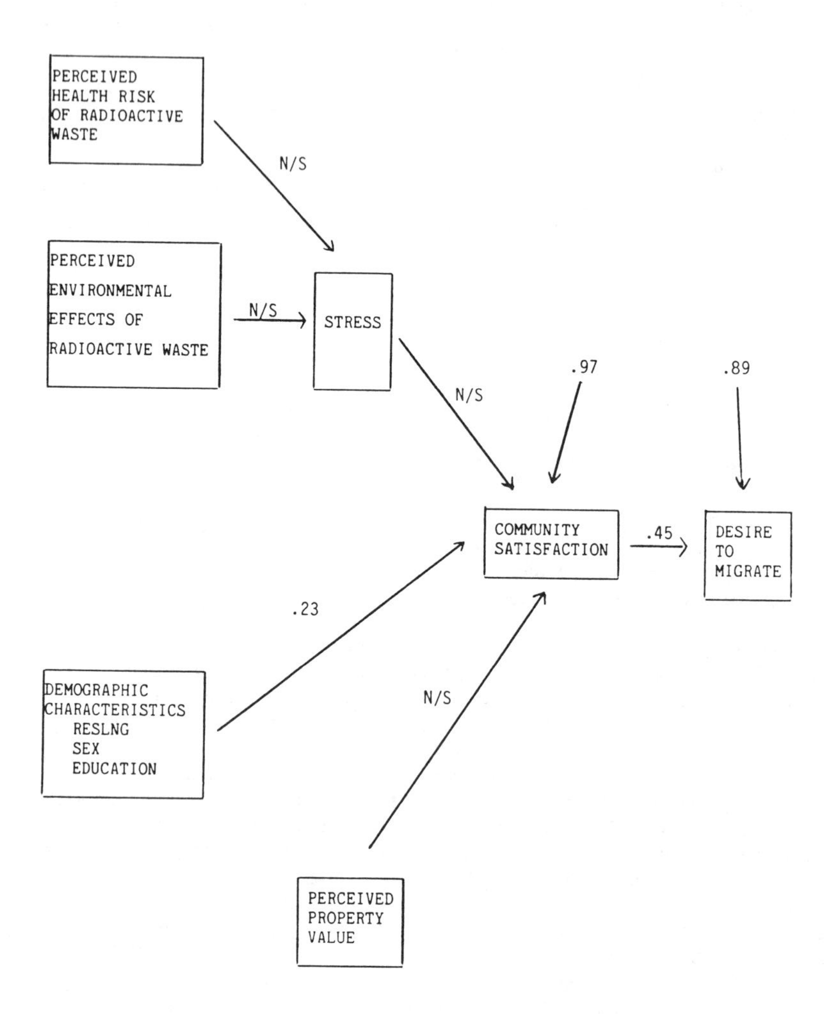

N/S = no significant relationship

and accounted for 20 percent of the variance in that variable. The standarized beta weights are shown in the path model for significant relationships along with the path coefficients from latent variables or residual causes.

Overall, the results of the analysis did not support the hypothesis that perception of hazardous wastes could affect one's evaluation of community. It was surprising, in fact, that 71 percent of the residents interviewed were satisfied with their community. Equally surprising was the finding that the waste site was only one factor among several that residents liked least about their community. In the case of West Chicago, it appears that individuals' perception of hazardous waste was not significant to define overall satisfaction (or dissatisfaction) with community.

Also surprising was the lack of relationship between perceived changes in property values and community satisfaction. Property values were studied two years earlier in this community with findings partially consistent with these findings [5].

A number of possible explanations for these findings exist. These include the possibility that important variables were not measured in this study; important variables were measured poorly; or the population sampled represented a subgroup that was not sensitive to the variables measured. It should be noted that community satisfaction and quality of life are ill defined, much debated concepts in sociology. As indicated in an article entitled "Dissatisfaction with Satisfaction" Sofranko and Fliegel note that global community satisfaction measures include more than is included in a standard list of community attributes [6]. These authors also note that the most important determinants of global satisfaction are "the perceived friendliness of neighbors --- the availability of outdoor recreation and shopping facilities". Thus, it is likely that global community satisfaction indicators may not be adequate for placing perceptions about hazardous waste in the broader community context.

An alternative explanation to the methodological problem noted above is the possibility that in communities where waste has been present for some length of time, residents dismiss the perceived hazard through social-psychological processes such as cognitive dissonance.

The results of this study help set the groundwork for future research on the social effects of hazardous waste. As suggested by some authors [7], the local "definition of the situation" may be important in understanding how people evaluate existing conditions in community life. Mileti and Williams [8] have suggested using the local definition of the situation as an important part of producing a sociological perspective on the siting of hazardous waste. Since social political problems are major concerns of waste managers, it seems reasonable to pursue scientific understanding of the way the waste effects community life. In the research reported here, waste did not influence perceptions as hypothesized. Additional basic sociological research will be needed to fully understand how waste affects perception and action.

References

1. Krannich, Richard S. and Thomas Grieder. "Personal Well-Being in Rapid Growth and Stable Communities: Multiple Indicators and Contrasting Results" Rural Sociology, Vol. 49, No. 4., Winter 1984.

2. Roht, L.H., W.W. Vernon, F.W. Weir, S.M. Pier, P. Sullivan and L.J. Reed. "Community Exposure to Hazardous Waste Disposal Sites: Assessing Reporting Bias" American Journal of Epidemiology, Vol. 122, No. 3, p. 418-433.

3. Knop, Edward. "Colorado Communities: A Decade of Change" (Survey Instrument) Colorado State University, 1982.

4. Norusis, Marija J. SPSS/PC+ for the IBM-PC/XT/AT Chicago, IL, 1986

5. Payne, B.A., S.J. Olshansky and T.E. Segal. "The Effects of Residential Property Values of Proximity to a Site Contaminated with Radioactive Waste" Waste Isolation in the U.S., Technical Programs and Public Education University of Arizona, Tucson, AZ, 1985.

6. Sofranko, Andrew J. and F.C. Fliegel. "Dissatisfaction with Satisfaction" Rural Sociology 49(3), 1984.

7. Deseran, Forest A. "Community Satisfaction as Definition of the Situation: Some Conceptual Issues". Rural Sociology 43 (Summer).

8. Mileti, D.S. and R.G. Williams. "A Sociological Perspective on the Siting of Hazardous Waste". Waste Isolation in the U.S., Techical Programs and Public Education, University of Arizona, Tucson, AZ, 1985.

REGIONAL HYDROGEOLOGICAL IMPLICATIONS ON THE PROPERTY TRANSFER ENVIRONMENTAL ASSESSMENT; A CASE STUDY

James Beck, ERT, A Resource Engineering Company

THE ENVIRONMENTAL ASSESSMENT

Long-term liability implications have evolved for past, current, and prospective property owners as a result of CERCLA legislation, and similar legislation that is increasingly occurring at the state level. For the buyer or seller in a real estate transaction, particularly in the commercial or industrial sector, it is becoming increasingly beneficial, if not a requirement in many instances, to obtain an environmental liability assessment of the parcel and/or facility under consideration. The assessment generally spans the entire spectrum of site history in an attempt to identify all activities that may have occurred at or near the location and that may have an impact on considerations concerning potential environmental impairment. The goal is to provide the buyer or seller sufficient information, at minimal cost, to protect his or her interest during and after closing of the actual transaction. To be effective, the assessment should be initiated as early-on as possible. That way minimal disruption occurs with respect to contract closing dates, etc. However, the time frame for acquisiton or divestiture of a complex commercial property should account for more in-depth investigative activity.

Reasons for the Assessment

The assessment is best described as an investigation into hidden conditions which might cause problems at a later date. These problems can be directly translated into financial liabilities on behalf of either the buyer or seller. The buyer's concerns are gener-

Geotechnical and Geohydrological Aspects of Waste Management, D. J. A. van Zyl et al., Eds.,

ally targeted toward establishing a degree of liability as a potential owner, whereas the seller can benefit from knowledge of existing environmental impairment, as he will always retain associated liabilities as a former owner. To determine the financial risk of exposure for either party, it is necessary to evaluate if potentially hazardous materials are on site, what migration pathways are available for human or environmental exposure, and what are the associated cost implications for introducing mitigative measures. Conversely, if mitigative measures are not considered viable (due to cost, time constraints, or whatever), it is necessary to provide a qualitative and quantitative assessment of risk or liability that could be borne by the legally responsible party. With a measure of the potential costs or liabilities, the buyer or seller has enough information in hand to assess his or her requirements relative to the transaction. The results of the assessment can therefore directly influence the terms and conditions of the sale.

Parts of an Assessment

An assessment involves investigatory effort performed by an environmental professional. Oftentimes, the environmental professional works directly on behalf of the client; however, increasingly he works alongside the client's legal representative to assure technically sound, legally defensible contingencies are structured into the contract of sale. The degree of confidence that can be attached to the final assertion is proportional to the level of effort expended, however at a minimum, the following topics are addressed.

- Historical use of the site
- Processes and substances used on the site
- Waste handling practices at the site
- Hydrogeology of the site
- Characteristics of adjacent properties
- Known environmental problems in the area
- Existence of obvious contaminant sources or areas indicating surface disturbance

All of the above are investigated to ascertain an initial indication of the extent, if any, to which the soil, groundwater, or surface water is contaminated as a result of current or past activities on site. This is accomplished through a variety of means, including:

- Interviews with site operations personnel or others familiar with the site
- Interviews with regulatory agencies
- Site reconnaissance
- Aerial maps, topographic maps

- Groundwater data, well logs
- Knowledge of processing methods
- Literature and records searches
- Permit status

Features that are involved in the site reconnaissance generally reveal considerable information on past practices at the site. Among the indicators are:

- Excavation or fill areas
- Discarded drums, other materials
- Impoundments, lagoons
- Discolored or stained soil
- Vegetative stress
- Wells, seeps, moist areas
- Underground utilities (migration pathways)
- Security features
- Adjacent land use

Determination of Findings

Assembling the information gained in the assessment will generally provide sufficient evidence or indications to enable an opinion on the degree of, or potential for, environmental impairment at the property. Oftentimes this opinion is rooted heavily in the experience and qualifications of the individual performing the assessment. It is sometimes found that the assessment effort has uncovered a need for additional investigatory effort (i.e., soil borings, observation wells, water quality analyses, etc.) to increase the level of confidence of the buyer or seller. The buyer/seller then has the flexibility to move forth with the additional effort (accompanied by the appropriate level of expenditure), or to make a decision with the information at hand, not having had to commit substantial sums to the venture.

The attractiveness of the environmental assessment, and the reason for its increasing and widespread application, is this ability to phase expenditure levels, thereby allowing the buyer or seller to maintain control at all times. As stated earlier, the assessment is an investment that can pay significant returns to either the buyer or seller. The following case study is exempletive of the type of information that can be uncovered in the course of a preliminary environmental assessment.

CASE STUDY

ERT was involved, in early 1986, in an assessment of a recently constructed warehouse facility site located in the southern portion of the City of Phoenix,

Arizona. Situated in a relatively new, planned industrial/office development, the facility is used to warehouse and distribute small aircraft engine components. A minor portion of the facility was used to repair some of these components, necessitating the use of small quantities of industrial solvents, such as Freon TF solvent and 1,1,1 Trichloroethane. Proper waste disposal procedures are adhered to and there was no visual evidence of any surficial contamination within the structure or confines of the grounds.

Ongoing regional groundwater monitoring programs have discovered elevated levels of solvent contaminants similar or identical to those in use at the facility. Arizona Department of Health (ADH) has proposed Superfund nomination for "area groundwater contamination" at an adjacent area, and is considering nomination of another area groundwater contamination site in the greater Phoenix area. This paper addresses the various hydrogeological inferences and potential liabilities that are of concern to the prospective purchaser of a facility that is potentially situated within the proposed boundaries of an "area groundwater contamination" site.

Site Description

The subject site (Figure 1) is situated immediately south of Interstate 10 (Maricopa Freeway), approximately 0.5 mile southwest of Sky Harbor Airport, in Maricopa County, Arizona. Consisting of over 24,000 m^2 (6 acres), it is occupied by a large [over 9,200 m^2 (100,000-square ft)] warehouse/distribution facility in the central portion of the property. With the exception of two small decorative areas near the site access point, the entire area surrounding the structure is overlaid with concrete or asphalt. The structure was built in 1983 and occupied approximately one year later by an aviation products firm.

Approximately 10 percent of the structure is dedicated to small component repair, with the remaining major portion being dedicated to warehouse/distribution activities. No actual manufacturing activity is carried out at this facility, and the component repair activities are restricted to one area in the structure. Most work stations are utilized for valve repair/disassembly that involves the use of solvents and produces a small quantity of waste fuel; the remainder are devoted to electrical component repair. Substances that were observed in use during the site inspection included Freon TF Solvent, 1,1,1- Trichloroethane, and Mobil jet lube oil. In the work area vicinity were drums (45.8 Imperial gallons) (55 gallons-U.S.)

containing waste fuel, used trichloroethane, and used trichlorofluoromethane. At each of the valve repair work stations there was a small solvent container (for parts cleaning) capable of holding 3.3 Imperial gallons (4 gallons-U.S.). All were intact, showing no signs of leakage or spillage. All waste streams (drums) are properly manifested to an approved disposal facility in California. The loading dock area where drums enter and exit the structure was observed to be void of any evidence of spillage. Waste metal materials are discarded into a secure vault/bin and sold for eventual meltdown and recycling.

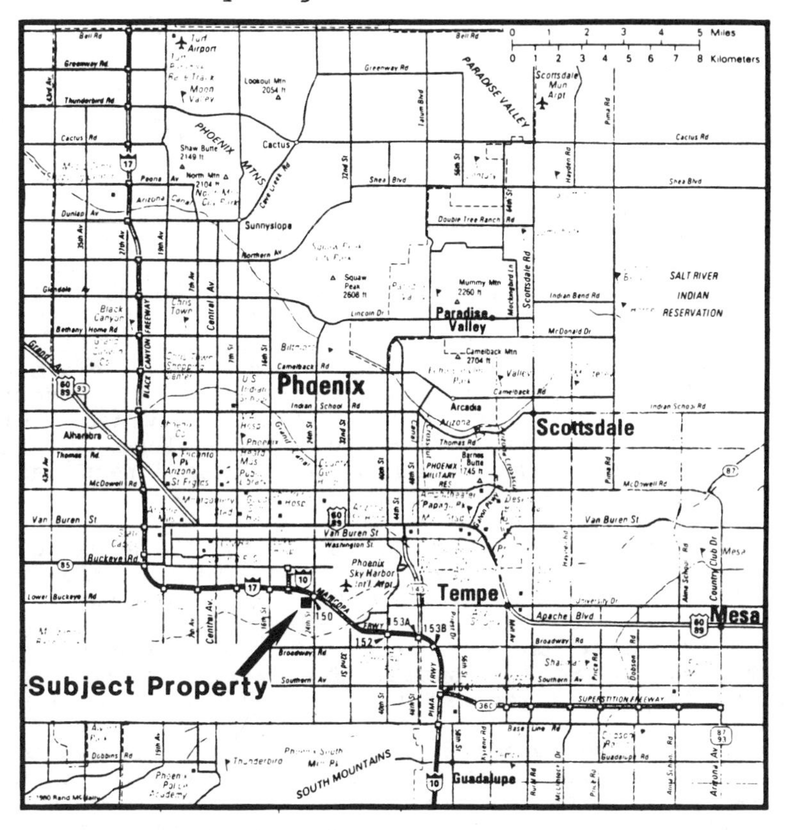

Figure 1: Site location.

The remainder of the structure was observed to be free of hazardous substances with the possible exception of a few occurrences of cleaning fluids, mineral spirits, etc. in locked, controlled cabinets. All areas where these substances were observed, including the work station area, were observed to be over competent concrete floor slabs.

The exterior perimeter of the structure consists of a concrete apron about 15 m (50 ft) wide on the west side, and 7.5 m (25 ft) wide on the south and east sides. The north side was buffered with about 9 m (30 ft) of decorative crushed granite restrained by a concrete curb enclosure. All sides then had asphalt overlays extending to the property boundary. The entire site was enclosed with "cyclone" fencing.

Neighboring facilities include a fashion distributor, office complex, a pool/spa distributorship, a custom truck parts shop, a wholesale auto parts distributor, a plastic bottle cap manufacturing plant, and a large retail distribution center. Most of the surrounding facilities present little concern, with the exception of two items of interest. The plastic bottle cap manufacturer had two bulk material silos located adjacent to the railroad track with a covered hopper standing on the spur. Additionally, the retail distribution facility had a vehicle refueling area located about 30 m (100 ft) from the subject property, apparently consisting of two underground storage tanks containing diesel fuel and gasoline. Another aboveground diesel tank, approximately 208 Imperial gallons (250 gallons- U.S.) was attached to the building as well. It should be noted that both of these sites are located downgradient with respect to surface drainage and "indicated" regional groundwater movement. Reportedly, the retail distribution firm subdivided their parcel to sell the acreage to the developer of the subject property and retained an ingress/egress easement through the property.

A railroad spur enters the property at the midpoint of the southern boundary. It curves in, parallel and adjacent to the structure and extends northward to the north end of the structure. An extension of the bottle cap manufacturer's spur parallels the southern boundary, ending at the access driveway. Neither spur facility is utilized by the occupant at this time.

All surface runoff from the site is channeled to a storm sewer system that is believed to discharge in the creek approximately 2 blocks west. The drainage eventually discharges into the Salt River (dry) bed after passing through several other facilities, including a large contruction company storage facility and headquarters.

Site History

Historically the subject property is known to have initially been a portion of the extensive sand and gravel mining operations that have been carried out in the Salt River floodplain since the 1930s. Reportedly, the gravel mining concerns have operated in a consistent manner over the years; excavating the raw material from pits [extending to 15-meter (50-foot) depths], advancing westward along the floodplain. As mining progressed, old pit areas were used as repositories for oversized, cobble material that was considered a waste product of the mining operation. Approximately two years ago (1984) the City of Phoenix commenced an in-depth investigation into past land uses (particularly landfill activities) along the Salt River floodplain, and determined that the mining/backfill process was carried out over an extensive area and that it was accompanied by randomized, indiscriminant dumping into the pits as well. Dumping was primarily road construction materials (concrete, asphalt, etc.); however, evidence was uncovered revealing household and industrial refuse. Reportedly, during the war years many aviation and electronics industries operated in the vicinity. As areas were completely backfilled and became surplus to mining operations, they were considered "reclaimed" and generally sold off to commercial development interests. As far as can be determined the subject property remained undeveloped until 1983, when the existing facility was constructed.

Discussion with the soils engineering firm that performed the soils investigation for the facility development, revealed that the entire property consisted of cobble/coarse material backfill. Corings and excavations revealed "concrete, asphalt, curb and gutter, manhole covers, and pipes". As such, construction specifications included excavation and replacement with new fill for the northernmost one-half of the property. It was not determined what the source of fill material was.

Available Hydrogeologic Data

Floodplain Status

The subject property, as shown in Figure 2, currently falls within the delineated boundaries of the 100-year (Salt River) flood hazard category. The Salt River bed is approximately 1.6 km (1 mile) wide near the subject property and approximately 1 km (0.6 mile) south at its centerline. According to the Arizona Department of Resources-Floodplain Management Division, the delineated area of the 100-year zone will decrease in the future due to improvement and upgrading of an extensive system of dikes or levies currently under construction along Salt River. This redesignation will take an estimated two years and require Federal

Emergency Management Agency (FEMA) approval. Reports indicate (City of Phoenix-Flood Management) that there were two 50-year and one 100-year floods all occurring in 1979. The flood insurance rating map (Figure 2) indicates a two-foot depth present one-block west of the subject property on December 4, 1979.

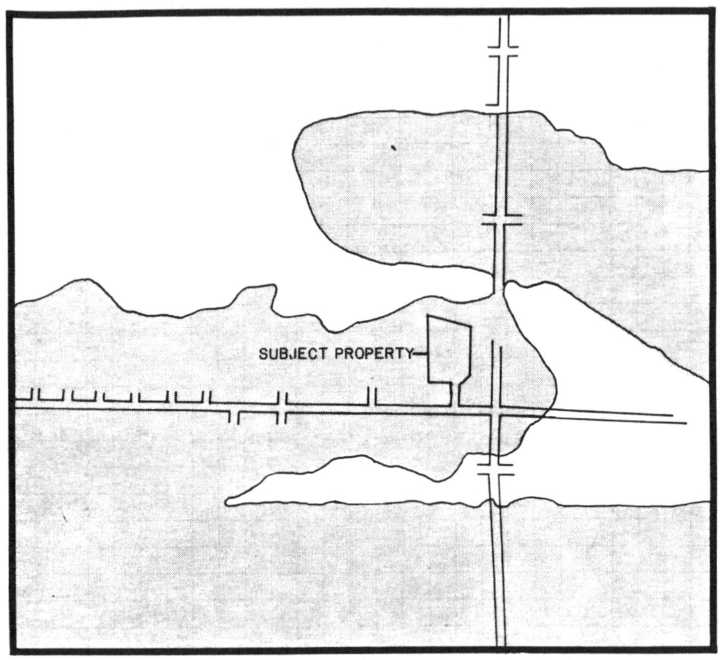

Figure 2: Location of subject within 100-yr. floodplain .

Groundwater Status

Site-specific groundwater data was not found to be available, however a regional study was recently completed by the Basic Data Office of the Department of Water Resources. Figure 3 (preliminary data, not yet finalized) indicates depth to groundwater, elevation of static (groundwater) level, and presents inferred static level isopleths. Groundwater movement in the subject property area is essentially westward at a depth to static water level of about 15 m (50 feet). Water quality data was obtained for a domestic well located about 3.6 km (2¼ miles) northwest of the site, and it revealed no abnormalities for tested parameters. It should be noted that the Salt River basin is comprised of sand/gravel/cobble alluvium that is considered quite homogeneous for the observed 15 m (50-ft) depth at the gravel mine. This type of material will exhibit high permeability and likely, relatively high rates of groundwater migration. The mining practice of backfilling pits with the oversize (coarse cobbles) should provide localized areas of proportionately higher permeability. As stated earlier, indications are that the subject property is situated in just such an area.

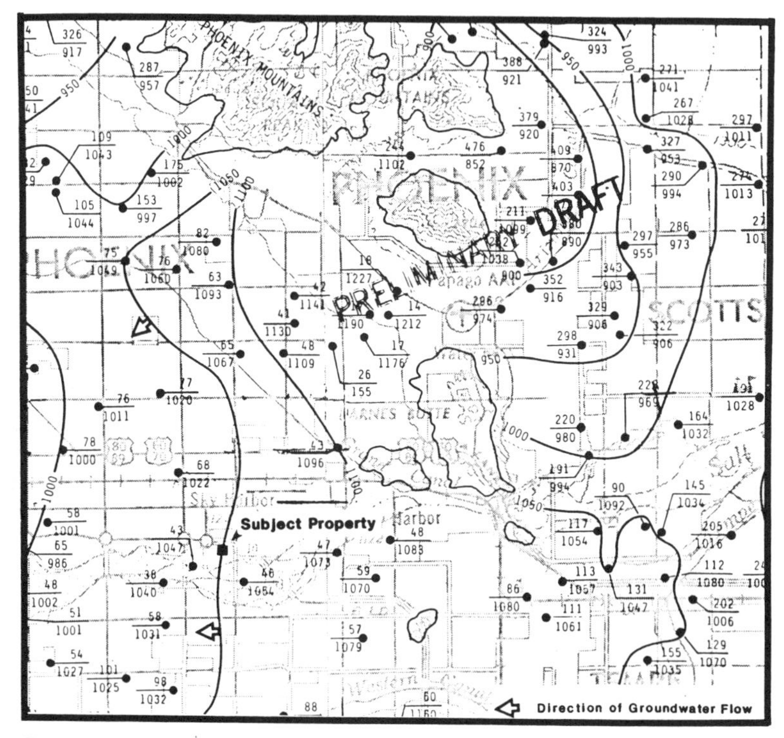

Figure 3: Regional groundwater characterization.

Current Site Status

There are several areas of interest regarding current site status. Capsulized summaries of investigative results are as follows.

Arizona Department of Transportation (ADOT)

Arizona Department of Transportation (ADOT) is currently constructing a storm runoff drainage tunnel about 0.4 km (.25 mile) west of the subject property as part of the Papago Project. The tunnel is described as having a 7.6 m (25-ft) outside diameter, concrete liner, and an invert elevation 16.8 m (55 ft) below grade. Due to the method being used to drive the tunnel and the characteristics of the overlying alluvium materials extensive dewatering and grouting techniques are being employed by the tunnel contractor. One of the dewatering well fields (ADOT #6) is located immediately behind the retail distribution center's employee parking lot (adjacent to the subject property). Production wells and observation wells were observed situated within a 61 km (200-ft) square area. Figure 4 shows the locations of the ADOT dewatering wells along with other monitoring stations in the area. Arizona Department of Health Services is utilizing the ADOT wells (and the other locations shown) as monitoring points to assemble data on the elevated CDE, TAC, TCE, and PCE levels in the vicinity. These substances, 1-1 dichloroethane; 1,1,1 trichloroethane; 1,1,2 trichloroethylene; and perchloroethylene are designated EPA Priority Pollutants. ADH has assembled a widespread geographic area (that includes the subject property) for nomination for inclusion under CERCLA/ Superfund listing. A hazard ranking score was developed but was not initially high enough to substantiate listing. Ranking considerations include degree of contamination and potential exposure to the public, such as through drinking water supplies. Because the area is not a major supplier of domestic/potable water, it scored low in the initial ranking. The ADH has considered a boundary expansion of the geographic area to include water supply source points at a further distance, thus enhancing the potential for a higher hazard ranking. As can be seen in Figure 4, the potential area is all-encompassing for the greater Phoenix area.

Rental Car Facility

A rental car facility, located near Mojave and 21st Street, reported a January 1985 underground gasoline storage tank leak of 10,800 Imperial gallons (13,000 gallons-U.S.). The contamination location is approximately .8 km (.5 mile) north of the subject property. Product has been detected in ADOT #7, but at this point in time has not been detected in ADOT #6. The approximate leak occurrence location is west of Sky Harbor Airport, near ADOT #7, shown on Figure 4.

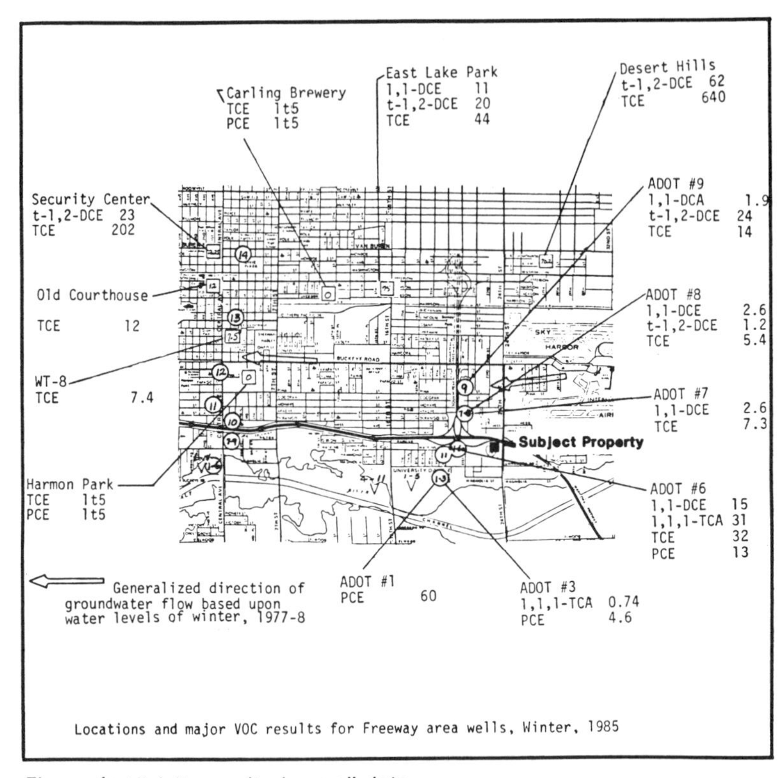

Figure 4: Vicinity monitoring well data.

Arizona Department of Transportation

Arizona Department of Transportation (State Highway Department) is preparing to widen Interstate 10 in the vicinity of the subject property. ADOT right-of-way extends to the subject property north boundary. It has been determined that in the area immediately north of the property, the ground (due to previous dumping in the area), is not sufficiently competent to stabilize the fill required for highway expansion. As such, ADOT has obtained a two-year temporary easement along the subject property north edge. This easement is to facilitate excavation activities wherein it is anticipated the entire underlying area will be excavated to approximately 3 to 4.6 km (10 to 15 ft) depth and backfilled with competent structural fill material. The State Highway Department's findings further quantify the reports of the site being used as a construction material landfill in the past. ADOT anticipates a need to remove and replace portions of pavement and parking areas on the subject site to assure adequate working area to provide a stable final embankment.

Conclusions

Considering the possibility of Superfund status at the site, the environmental liability potential of the subject property would have to be considered as medium to high. The fact that the occupant uses some of the same substances as those detected, (even though they may be operating in a responsible manner) would have a tendency to draw attention to this property and could potentially trigger an extensive site investigation for the owner of record. The potential purchaser should be cognizant of the legal and financial ramifications of being included in what could become a Superfund site before moving forth with the transaction.

REFERENCES

1. Arizona Department of Health Services.

2. Basic Data Office, Arizona Department of Water Resources.

3. City of Phoenix, Engineering.

4. Papago Coordinating Office, City of Phoenix.

5. City of Phoenix, Solid Waste Disposal.

6. Western Technologies, Inc., Consulting Soils Engineers Phoenix, AZ.

7. Arizona Department of Transportation.

8. William A. Duvell, Jr., Ph.D., P.E., "Environmental Termite Inspections," Hazardous Waste Management for the 80's, Ohio State University, 1982.

MANAGEMENT FOR HAZARDOUS WASTE LIABILITY AT MINING SITES

William E. Cobb and W.V. Bluck, CH2M Hill, Denver Colorado

Ian P.G. Hutchison, Steffen Robertson and Kirsten, Denver, Colorado

INTRODUCTION

Mining companies, as developers of natural resources, are being forced to deal with potential long-term liabilities of past, present, and future mining operations. These liabilities are very real, and take several forms: RCRA compliance and cleanup; Superfund; or Natural Resource Damage Claims (NRDC). This paper discusses management strategies for dealing with these various liabilities. The authors' observations and suggestions are based on corporate experience with approximately 15 mining-related Superfund, over one hundred organic-contaminated Superfund, and fifty private-sector hazardous waste projects.

The financial liabilities associated with mining-related Superfund sites and NRDC's are becoming evident; potential financial liabilities are demonstrated as follows:

- Union Carbide recently settled on a $42 million cleanup program at Uravan, including moving processing wastes to centralized storage areas; and
- State of Colorado Department of Health has proposed a cleanup of $120-200 million for the Eagle Mine, including tailings removal to a RCRA disposal facility and perpetual treatment of mine water.

As evidenced by these numbers, the potential financial impacts of mining-related environmental liabilities can be extremely high. Settlements could severely impact the profitability of any company, whether it is large or small.

PROBLEM DEFINITION

Waste management problems can result in burdensome liabilities and legal entanglements. Extreme cases include the mine site

Geotechnical and Geohydrological Aspects of Waste Management, D. J. A. van Zyl et al., Eds., © 1987 Lewis Publishers, Inc., Chelsea, Michigan – Printed in USA.

being placed on the National Priorities List (NPL), (i.e., Superfund) and/or subjected to a NRDC.

The Comprehensive Environmental Response, Compensation, and Liability Act of 1980 (CERCLA) created a mechanism for investigating and cleaning up the top priority hazardous waste sites in the United States. The five-year, $1.8 billion, program initiated studies and cleanup programs at numerous facilities. As popularly claimed, the Environmental Protection Agency (EPA) has to date only cleaned up a small number of sites. However, this statement is extremely misleading; it is difficult to delist (cleanup) a site if the Record of Decision requires 30 years of monitoring to assess the success of remediation. One would, therefore, expect only a small number of sites to have been delisted after five years.

The Superfund Amendments and Reauthorization Act of 1986 (SARA) provides approximately $9 billion for another five years of Superfund. Besides continuing the original Superfund process, SARA includes the following provisions;

- A minimum number of site investigations and site cleanups is specified per year, resulting in a more concentrated effort to complete site studies and implement remedial solutions;
- Maximum Concentration Limits (MCL's), based on enforceable standards for a given media (for example, groundwater or surface water), will be applied as cleanup criteria when evaluating remedial alternatives;
- The philosophy that state cleanup standards can be more stringent than federal cleanup standards is continued;
- Permanent site remedies will be investigated and selected, where possible and practical. Permanent solutions, such as incineration of organic wastes or reprocessing of inorganic wastes, are preferred to waste removal and land disposal. Landfill storage is viewed as potentially moving the problem from location A to location B; and
- Innovative and emerging technologies will be investigated and selected, where possible and practical. Less importance may be placed on selecting proven technologies if emerging technologies can provide permanent solutions to hazardous waste problems.

Among the least chartered waters of CERCLA, and now SARA, are the provisions relating to NRDC's. CERCLA defines natural resources as:

- "Land, fish, biota, wildlife, air, water, groundwater, drinking water supplies, and other such resources belonging to, managed by, held in trust by, appertaining to, or otherwise controlled by the United States......any state or local government, or any foreign government."

CERCLA provides a mechanism for the local, state, or federal government to recover damages to these various resources via litigation against potentially responsible parties (PRP's). These resource damage assessments are above and beyond the cost of remedial action, and can range up to $50 million in liability.

NRDC's can be potentially onerous to mining operations and companies. The government (either local, state, or federal) can file a NRDC if significant environmental damage has occurred as a result of the mining operation. The site does not have to be on the NPL or proposed NPL to have a NRDC filed against it. A good example is the Eagle Mine, previously mentioned, which was not listed on the NPL at the time of the draft Record of Decision.

NRDC's have yet to become prevalent; several states (including Colorado, Idaho, and Utah) have filed claims relating to mining activities. State agencies have tended to file NRDC's. Proposed remedial actions typically have been predicated on RCRA cleanup standards. These standards generally require cleanup to background (in the case of soils) or MCL's, in the case of water. Some recent examples, from the Eagle Mine, of RCRA level cleanup include:

- Removal of 8 million tons of tailings to a RCRA-approved disposal facility (the nearest is in western Utah); and
- Permanent pump and treatment facility to clean up mine waste (acid-mine-drainage) to MCL's (State Drinking Water Standards).

With a possible change in the Hazardous Ranking System, it is likely that only a few mine sites will be added to the NPL. However, in the longer term, mining companies will probably see increased numbers of NRDC's. The NRDC approach offers the states and local governments a relatively straightforward mechanism to pursue a range of environmental and human health improvements. For example, a State NRDC could be used to force Company X to cleanup a non-point discharge, such as groundwater contamination migrating from a tailings impoundment.

The Federal and State governments, as documented by current Superfund and NRDC actions, are serious in their determination to remediate hazardous waste sites. These cases, as evidenced in the introductory statements, can be extremely expensive because of both technical and legal costs. These issues, therefore, require that: (1) mining companies employ the best possible resources to deal with waste management problems; and (2) adopt an enlightened and aggressive approach to grappling with the issues.

WASTE MANAGEMENT STRATEGY

Faced with these potentially onerous liabilities, the mining companies would be wise to develop and implement a sound waste management strategy for new mines coming on line, as well as existing operations, and non-operating closed mines. A sound waste management strategy can provide a cost savings of up to one order of magnitude at some sites. This strategy must be based on having a solid technical defense against the potentially obtuse legal positions of a regulatory agency. CERCLA and RCRA, although potentially onerous, can be confronted using:

- A superior technical approach, relying on Risk Assessment techniques to guide problem-solving and decision-making;
- A solid understanding of the regulations and their

applicability to problems found at a given site;

- An aggressive negotiating posture with the regulatory agencies, using technical personnel rather than lawyers. Furthermore, the company must be prepared to back their negotiations up with the implementation of their proposed remedial actions; and
- The application of a staged approach to the engineering and implementation of remedial actions can, in many cases be used, to minimize front-end costs and to avoid installation of redundant remedial actions.

These elements of a sound waste management strategy are discussed below in more detail:

1. A superior technical approach.

The CERCLA and NRDC decision-making system is based on using engineering judgement to evaluate potential remedial actions that are compatible with site conditions and meet the cleanup goals and objectives. Because there are many ways of decreasing contaminant levels, technical innovation and judgement are crucial to the waste management strategy. If a regulatory agency believes a company's technical arguments, the company's proposed remedial actions will probably be approved.

Environmental and health problems are typically associated with tailings impoundments, waste rock piles, stockpiles, slag/processing wastes, and mine drainage. Technically innovative and cost-effective solutions to these potential pollution problems exist; however, the solutions are easier to implement for new ventures than to retrofit during site closures:

- Surface solids and water can interact to form suspended sediments and metal-laden drainage problems. Surface water diversion, waste surface management, and drainage collection technologies can be used to keep contaminants away from surface waters. Barriers, and in some cases natural geologic formations, can be used to keep contaminants away from groundwater systems. Based on site conditions and waste characteristics, application of several of these technologies can provide a cost-effective means of protecting public health, welfare, and the environment (PHWE); and
- Acid-mine-drainage (AMD) can emanate from both surface and underground mines. AMD can potentially impact both groundwater and surface water systems. Collection and treatment technologies, although sometimes perpetual in nature, can minimize the impact. Barrier and sealing technologies, depending on site conditions, could decrease AMD flows. Combinations of these technologies can provide cost-effective means of protecting public health, welfare, and the environment.

Secondly, the company must be able to communicate using the jargon of the regulatory agency. Agencies generally use a Risk Assessment to determine if these sources or potentially contaminated media pose a threat to the PHWE. Demonstrating a threat to the PHWE, using Risk Assessment allows EPA or the State to pursue Superfund actions; Risk Assessment is also used as a means of delisting a site. It is, therefore, important for mining companies to evaluate their own site problems, and to use Risk Assessment technology to quantify potential problems. If problems are identified, the technical team can use Risk Assessment to focus selection of remedial actions on the important issues and to formulate presentations to the regulatory agencies.

2. A solid understanding of applicable regulations

   There are many regulations that may be applicable to problems found at a given site. These include: drinking water standards, groundwater standards, ambient water quality for aquatic life, air quality standards, etc. These standards address both human health and welfare, and environmental impacts. Some of these standards are enforceable, and some are non-enforceable. Cleanup criteria can be either MCL's or Alternate Concentration Limits (ACL's). ACL's would be higher than MCL's, and typically represent the achieveable cleanup level taking into consideration conditions or technology.

   A thorough understanding of these regulations is required to successfully negotiate with agencies. By knowing the options available in the regulations, an astute executive can potentially save his company significant sums of money by negotiating the acceptance of the most cost-effective site remedy.

3. An aggressive negotiation posture

   A company has basically two choices on how to proceed in the event that their Risk Assessment indicates site environmental damage or health risk problems. It can wait until an agency does something or it can aggressively proceed to remedy the situation. The authors' experience indicates that the aggressive method tends to lower site investigation, as well as remediation costs. This experience also indicates that, in general, technical people, problem-solvers, should undertake negotiations with agencies. By using superior technical knowledge of site conditions and technologies, the company can potentially diffuse the possible legal objections to proposed remedial actions. The sit-and-wait attitude tends to result in extensive battles in court, with attorneys often dictating site investigation and remediation costs. However, there are always exceptions to the rule; executive judgement should be used to detemine the extent that legal support is required for agency negotiations.

   The company must be prepared to put its money where its mouth is, and must be prepared to implement its proposed remedial or correction actions. This part of the strategy is directly linked to superior technical knowledge. Confidence in one's ability to implement a solution and meet performance goals will probably lead to agency acceptance of the proposed concept.

4. A staged approach
The company should use a staged approach to decision-making and implementation of remediation at an existing site. This process is based on ranking the pollution problems at the site, followed by the development of sequentially staged remedial alternatives that initially deal only with the major problems. Installation of the remediation would follow this sequence. This facilitates field verification of the projected beneficial impacts. Furthermore, adjustments or in some cases curtailment to subsequent stages of the program, can be continually made. In this manner, the company is able to avoid the inclination to study and evaluate everything in detail. Large, unneeded, expenses can be prevented, and cost-effective studies and remedial programs can be achieved.

CONCLUSIONS

- We can expect renewed CERCLA activity with the approval of SARA;
- Superfund remediation and Natural Resource Damage Claims are very serious and potentially extremely costly liabilities;
- Mining companies should place significant emphasis on implementing comprehensive, scientifically based, waste management strategies for new, ongoing, and non-operating facilities; and
- A sound waste management strategy should be based on the following components:
  - A sound technical understanding of site conditions and contaminant migration pathways;
  - A solid understanding of Risk Assessment techniques, identifying potential site problems that could impact public health, welfare, and the environment;
  - A sound grasp of regulations that may be applicable (both enforceable and non-enforceable) at a given site;
  - A superior technical understanding of the implementability of specific, cost-effective, remedial or corrective actions;
  - An aggressive negotiation approach that uses technical personnel rather than attornies;
  - The willingness to implement negotiated remedies in a timely manner; and
  - A staged approach to the implementation of remedial actions.

PERMEABILITY STUDY OF LANDFILL SOIL LINER

R. Janardhanam, The University of North Carolina at Charlotte, Charlotte, North Carolina

A. Perumal, Johnson C. Smith University, Charlotte, North Carolina

ABSTRACT

Leaking hazardous waste landfills present a potential contamination problem to adjacent groundwater aquifers. This investigation selects a good soil liner for the State of North Carolina and examines its performance conditions under which the soil liners would be exposed in the field are replicated. A triaxial pressure permeameter is developed to simulate the field conditions on the sample. The effects of chemical leachates on soil liner's permeability characteristics and filtering capacity are investigated.

INTRODUCTION

The land disposal of hazardous waste materials in North Carolina is at present a problem of great concern. The fear for contamination of groundwater due to leaking landfills causes serious public concern. Furthermore, the dependence of this country and this state in particular makes the protection of the groundwater supply from contamination imperative. Also, the purification of a contaminated groundwater aquifer is an unrealistic remedy. Therefore, a strong need exists to evaluate and design stable soil liners.

This study involves the selection of a soil liner and testing its suitability, compatability and stability. It is tested with a typical leachate placed

Geotechnical and Geohydrological Aspects of Waste Management, D. J. A. van Zyl et al., Eds., © 1987 Lewis Publishers, Inc., Chelsea, Michigan – Printed in USA.

or generated in hazardous waste landfills in this state. The study is primarily focused to study the physical and engineering properties of natural soil and Bentonite blended soil and test them for their physical and chemical properties, their permeability characteristics, and their compatability with chemical leachate.

BACKGROUND

North Carolina State, in general, has very unfavorable geologic conditions for hazardous waste disposal by means of landfills due to the ease of leachate migration from the fills. The Coastal Plains are invariably made up of sands and gravels and the Piedmont area is underlain by crystalline rocks. The surface soil in this area and Sarpolite are fairly pervious. Groundwater aquifers are also not deep. Shallow aquifers are easily accessible and widely used for groundwater supplies; they are susceptible to contamination due to leachate migration. In addition, the number of housing units using onsite water supply systems is increasing.

The unfavorable geologic conditions and the growing demand for clean and quality groundwater encourage us to find the ways and means to protect the groundwater aquifers. Proper lining of the new landfills may help to reduce the potential for contamination.

The liner is the most critical part of a landfill. It should be designed to minimize leakage. A liner is expected to impede the flow of pollutants and pollutant carriers and absorb suspended or dissolved pollutants. Liner made of compacted local soil or local soil blended with proper admixture serve this function well. Addition of admixture is often necessary to improve the permeability characteristics of local soil to have the desired properties as recommended by the Environmental Protection Agency (EPA).

When soil liner is exposed to a wide spectrum of chemicals, its permeability characteristics are observed to be adversely affected. There are structural changes such as volume reduction, development of shrinkage cracks and chemical changes such as dissolution of carbonates and sulphates. The performance of a soil liner and the mechanism of interaction between the soil liner and chemical leachate is the subject of study in this paper.

OBJECTIVE

The objectives of this study are (1) to study local soil as liner for a landfill, (2) to test the performance of local soil blended with admixtures, and (3) to study the performance of selected liner mix for chemical resistance and permeability.

A triaxial pressure permeameter is developed to simulate the true field conditions. This study with a chemical leachate as a permeating fluid may aid in understanding the physico-chemical processes which occur within the soil liner.

SOIL SELECTION

The local soil used as a liner material in landfills is tested in this investigation. The soil is brick red in color and classified as silty clay (OH) according to Unified Soil Classification system. The coefficient of permeability (K) of the soil is 5.01 X 10^{-5} cm/sec at its maximum proctor compacted density of 92.5 pcf. The optimum moisture content is 21%. The clay mineral contained in the soil medium is identified by base exchange capacity to be Notreonite.

MIX DESIGN

EPA regulations require hazardous waste landfills and surface impoundments to have liners with permeabilities (K) not to exceed 10^{-7} cm/sec. As the local soil has higher value of K, different percentages of admixtures (here Bentonite clay) are mixed and tested to establish a mix to possess a K value of 10^{-7} cm/sec or less. Bentonite used in this investigation is a sodium rich montmorillonite and is commercially available. In selecting the mix proportions, three factors, namely maximum dry density, swelling potential, and coefficient of permeability, are considered as deciding factors. The mix design selected is then tested for chemical resistance.

Details of the test procedures and the results are presented in the following sections.

CONSISTENCY LIMITS

The consistency limits of the various soil-bentonite mixes are presented in Table 1. As the percentage of Bentonite increases, the specific gravity decreases, but the liquid limit increases. Though the shrinkage limit decreases, the shrinkage ratio increases. The swelling of bentonite upon absorption of water is an advantage when using it as a sealant in liners. But its volume reduction upon exposure to chemical leachates causes great concern.

The Lattice structure of the montmorillonites is a weakly bonded one. The interlayer spacing changes upon inhibition of water. The interlayers are initially occupied by loosely held water and exchangeable metallic ions. Montmorillonites exhibit the greatest cation exchange capacity. It can absorb over 200% of its solid phase weight and consequently has the capacity for large shrinkage if its interlayer water is displaced by other fluids that yield lower interlayer spacing.

MOISTURE DENSITY RELATIONSHIP

Proctor compaction tests have been conducted to study the moisture density relationship of the various soil-Bentonite mixes (ASTM D 698). The results are shown in Figure 1. The maximum dry density and optimum moisture content both increase with the addition of Bentonite up to 2% and then drops. Therefore, 2% Bentonite-soil mix is considered to be the optimum blend that gives the most dense soil liner.

SWELL POTENTIAL

To study the swelling potential of local soil and soil-Bentonite mixes, free-swell tests have been conducted. In free-swell test, Figure 2, the soil-2% Bentonite mix shows the maximum swelling. The mixes with greater percentages of Bentonite did not swell appreciably, probably the swollen Bentonite acted as a blanket and prevented the unaffected Bentonite from getting affected. Whereas in confined swell, Figure 3, the swelling is seen to increase with bentonite content. However, there is a significant jump in swell potential of the mixes from 1% to 2% bentonite and thereafter the difference narrows down. These two test results indicate in both instances that the 2% bentonite-soil has enough swell potential to be used as a fill liner.

PERMEABILITY TEST

Falling head permeability tests have been conducted on compacted soil and soil-Bentonite samples. The specimens have been prepared at the Proctor's maximum dry density, then fully saturated by "back-suction" technique. The fully saturated samples are then tested for their permeability characteristics. The test results, Figure 4, show a monotonic increase in K value with increase in percentage of Bentonite. To achieve a K value of 10^{-7} cm/sec, 1.5% of Bentonite blend is found to be sufficient. However, 2% Bentonite soil blend is chosen as it has the maximum dry density and also maximum swell potential.

TRIAXIAL PRESSURE PERMEAMETER

The pressures (stresses) acting on an infinite small element of soil liner is shown in Figure 5. It is subjected to vertical stress σ_1 and a lateral stress σ_3. The vertical stress is essentially due to overburden material on the liner. In addition to this overburden pressure, fluid pressure generated by the impoundment of liquid waste in the fill also acts in the vertical direction. Under such environment only, the soil liner generally exists. A realistic study on the performance of soil liner in the laboratory would be possible if only the true field conditions are replicated. The triaxial pressure permeameter designed and devloped successfully simulates the field conditions well.

Figure 6 shows the principles involved in the design of triaxial pressure permeameter. Compressed air is used to apply the confining pressure σ_3 to the sample. The fluid pressure is introduced by inleting the permeant to flow through plunger and porous disc. The overburden pressure σ_1 (= σ_3 + Δp) is applied through the axial shaft. The fluid passes through the soil sample, trickles down through the outlet and is collected. The permeameter is used to study the behavior of the soil liner with 0% and 2% Bentonite when exposed to distilled water and pesticide liquid cabaryl. Samples prepared with three different void ratios (compaction efforts) are tested.

PERMEABILITY ANALYSIS

Tests on compacted specimens of local soil reveal that smaller the void ratio, lower will be the value of K. Soil permeates more when exposed to a chemical fluid cabaryl. K value increases by about 200 times as shown in Figure 7. However the influence of void ratio on K remains at the same rate. When 2% Bentonite blend soil specimen is tested the loss of K value is about 55% only as shown in Figure 8. The reduced value of K (= 3 X 10^{-8} cm/sec) is still far more than the EPA's requirement of 10^{-7} cm/sec. Therefore, addition of Bentonite adequately improves the permeability characteristics even under adverse environments.

POLLUTANT MIGRATION ANALYSIS

A U-V spectrophotometer is used to determine the concentration of cabaryl permeating through the soil specimen. The analysis shows that it takes more time for the concentration output to equal that of the input, in case of 2% Bentonite blend soil than in the native soil, shown in Figure 9. The negative log of transmission when plotted against wave length shows a smooth transition for 2% Bentonite blend soil in Figure 10. The filtering capacity is seen to be better with the blended mixture.

DIFFRACTION ANALYSIS

Failure of soil liners can be characterized by cracking, internal erosion, soil piping, structural changes and clay dissolution. Some of these are due to the change in water content that could result from either the extraction of interlayer water or displacement of interlayer water. This suggests that in order to remain impervious, a soil liner should not shrink upon loss of water content and should not permit the formation of seepage paths. Preliminary X-ray diffraction study indicates, clearly, the possibility of following structural changes in the blended samples during hydration-dehydration cycles shown in Figure 11. A quantitative assessment of dilation on the molecular level is currently underway using the diffraction measures.

CONCLUSIONS

Soil liners made of compacted local soil blended with specific quantities of Bentonite are tested. The native soil is identified to be silty clay containing the clay mineral Notronite and has a low coefficient of permeability. The optimum percentage of Bentonite to be blended with the local soil to get the desired permeability characteristics is determined to be 2% by laboratory investigation. A truly triaxial pressure permeameter designed to simulate the field conditions gives reproducable results. The native soil loses its imperviousness significantly upon exposure to a chemical liquid, cabaryl. Whereas the 2% Bentonite blended soil misture has shown appreciable resistance to reduction of K value. The filtering capacity is also seen improved by the addition of Bentonite. The Bentonite blend makes the transmission of pollutants through soil medium slow and smooth.

ACKNOWLEDGEMENT

Authors wish to express their appreciation to Dr. Louis G. Daignault, Mr. George D. Barrier, and Mr. Gilberto Ramos for their participation in this study. The financial support through the UNCC and Johnson C. Smith University Faculty Summer Research Grant 1985 are gratefully acknowledged.

REFERENCES

1. Anderson, David. "Does Landfill Leachate Make Clay Liners More Permeable?" ASCE Civil Engineering Journal, September, 1982.

2. Anderson, David, Brown, K. W., and Green. "Effect of Organic Fluids on the Permeability of Clay Soil Liners," Land Disposal of Hazardous Waste, Proceedings of the Eighth Annual Research Symposium (EPA-600/9-82-002), January, 1982.

3. Barrier, George D. and R. Janardhanam. "Compatability of Soil Liners in Landfills," Collegiate Academy of North Carolina of Science, Publication No. 1, Volume 30.

4. Bolt, G. H. Soil Chemistry. Elsevier Science Publishing Company, Inc., New York, NY, 1982.

5. Grim, R. E. Clay Mineralology. McGraw-Hill Book Co., Inc., 1953.

6. Haxo, H. E. "Durability of Liner Materials for Hazardous Waste Disposal Facilities," Land Disposal of Hazardous Waste, Proceedings of the Seventh Annual Research Symposium (EPA-600-81-002), 1982.

7. Morrison, Allen. "EPA's New Land Disposal Rules - A Closer Look," ASCE Civil Engineering Journal, January, 1983.

8. Morrison, Allen. "Can Clay Liners Prevent Migration of Toxic Leachate?" ASCE Civil Engineering Journal, July, 1981.

Table 1

Index Properties of Soils

Test sample		1	2	3	4
Soil	Bentonite (%)	0	1	2	3
Blend	Soil (%)	100	99	98	97
Specific Gravity		2.76	2.74	2.66	2.64
Liquid Limit		60	65	80	77
Plastic Limit		47.5	37.3	36.5	33.5
Plasticity Index		12.5	27.7	43.5	43.5
Shrinkage Limit		39.6	34.7	31.3	27.6
Shrinkage Ratio		1.45	1.53	1.55	1.57

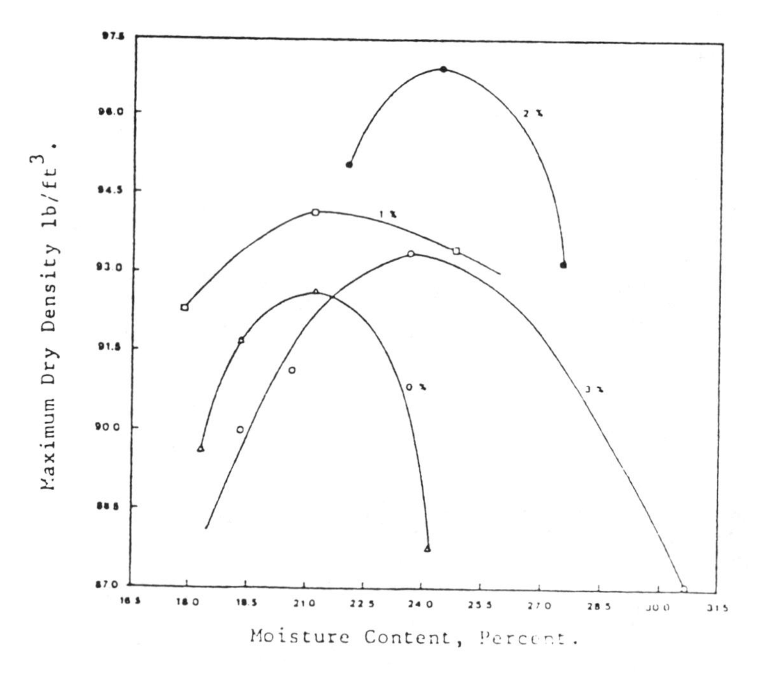

Figure 1. Moisture - Density Relationship

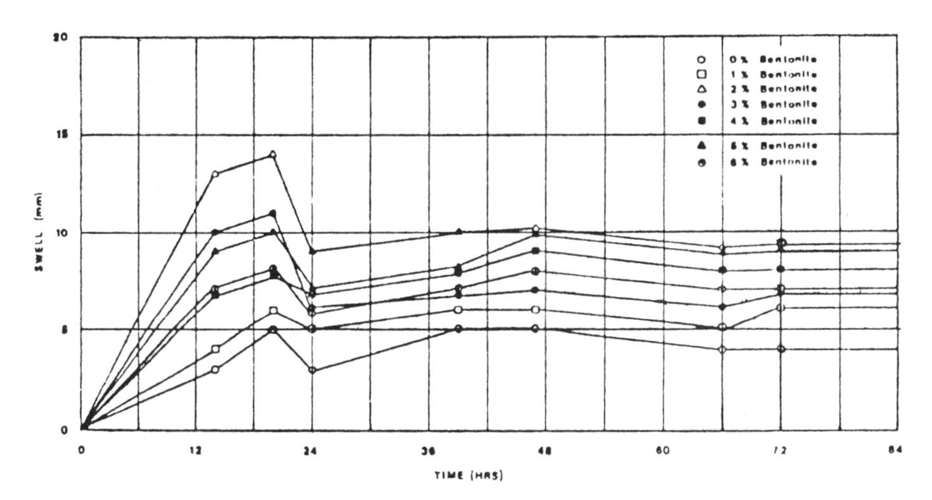

Figure 2. Free-Swell Potential of Soil-Bentonite Mixes.

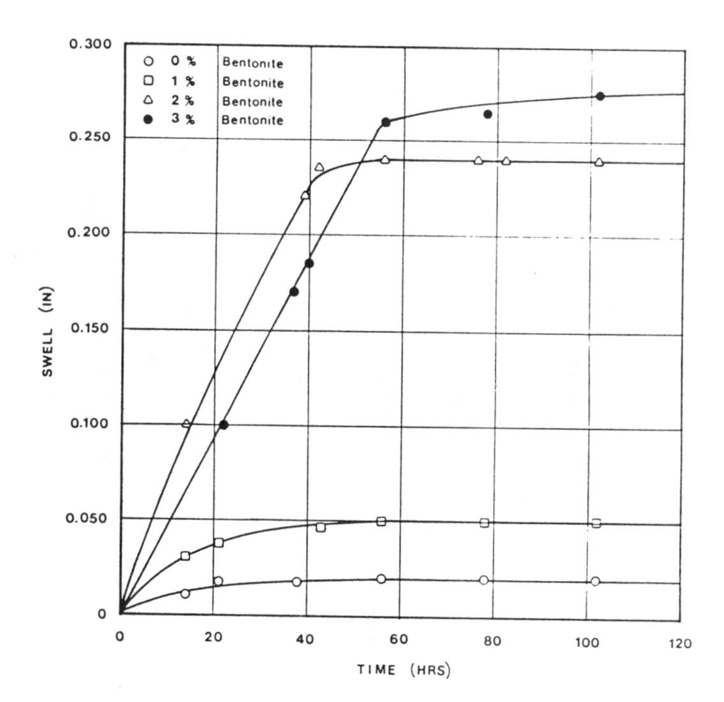

Figure 3. Confined - Swell Potential of Soil-Bentonite Mixes

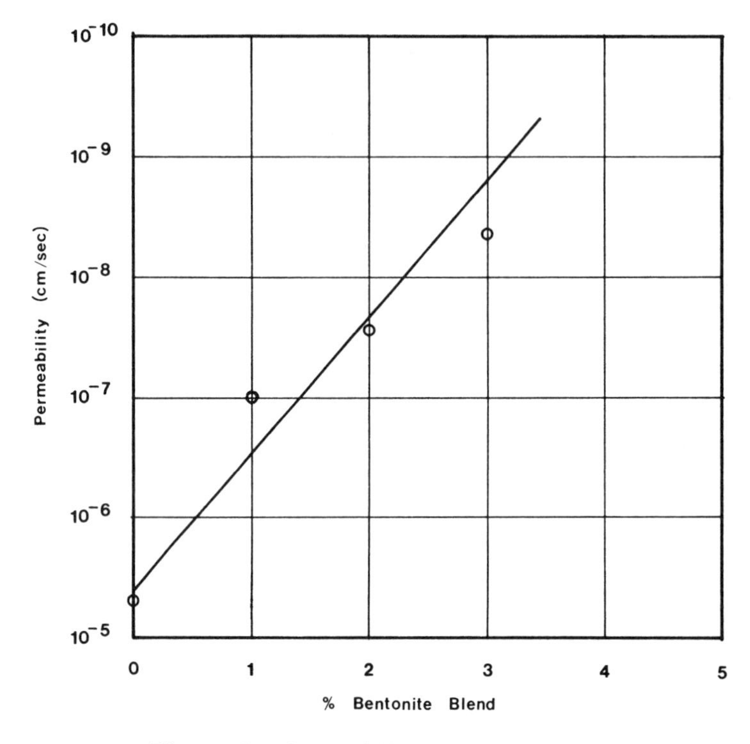

Figure 4. Permeability of Soil - Bentonite Blend Mixes.

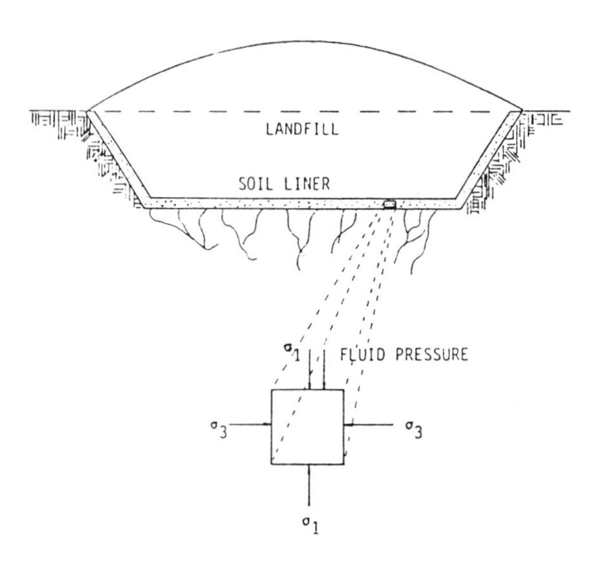

Figure 5. Soil Under Stress - In a Landfill Liner

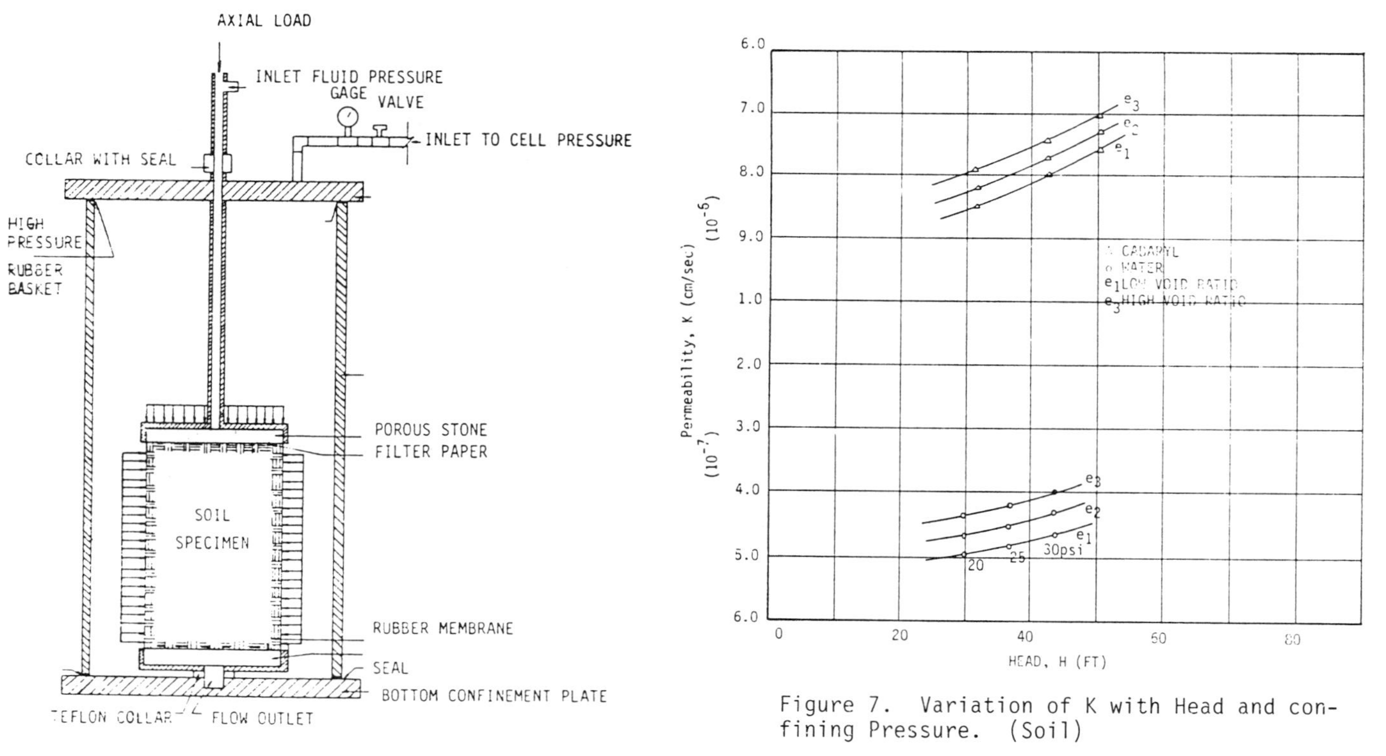

Figure 6. Triaxial Pressure Permeameter

Figure 7. Variation of K with Head and confining Pressure. (Soil)

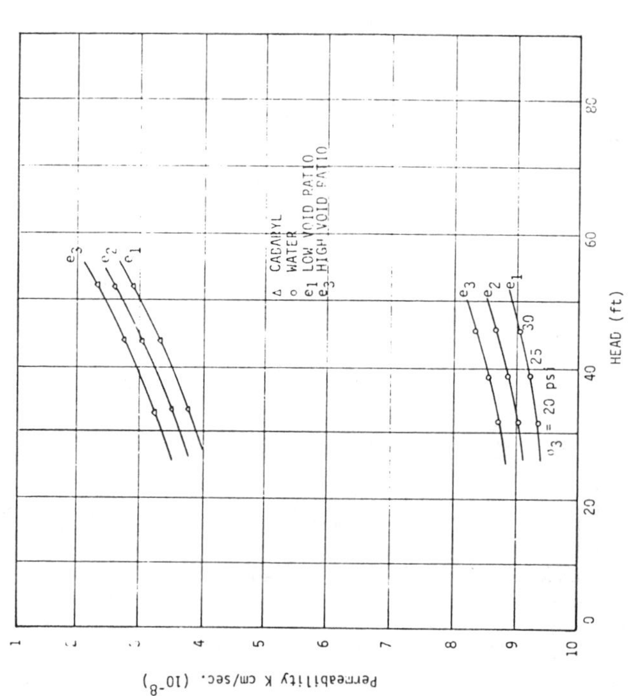

Figure 8. Variation of K with Head and confining Pressure. (Bentonite Blend Soil)

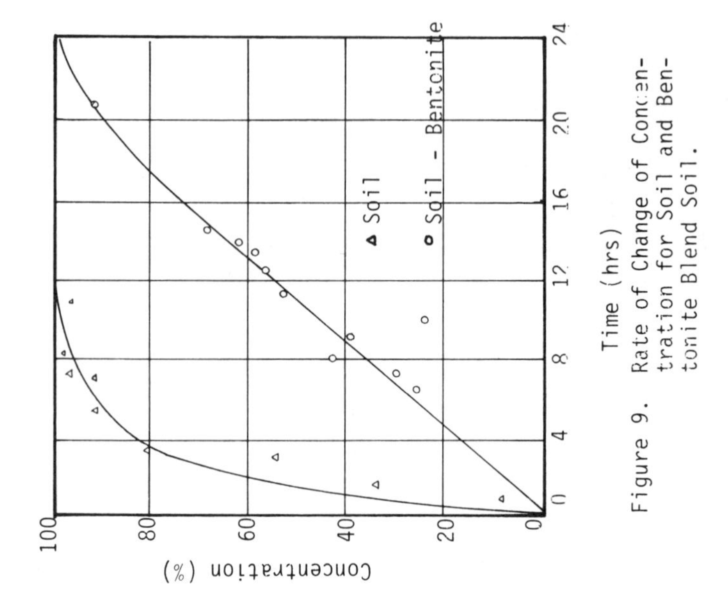

Figure 9. Rate of Change of Concentration for Soil and Bentonite Blend Soil.

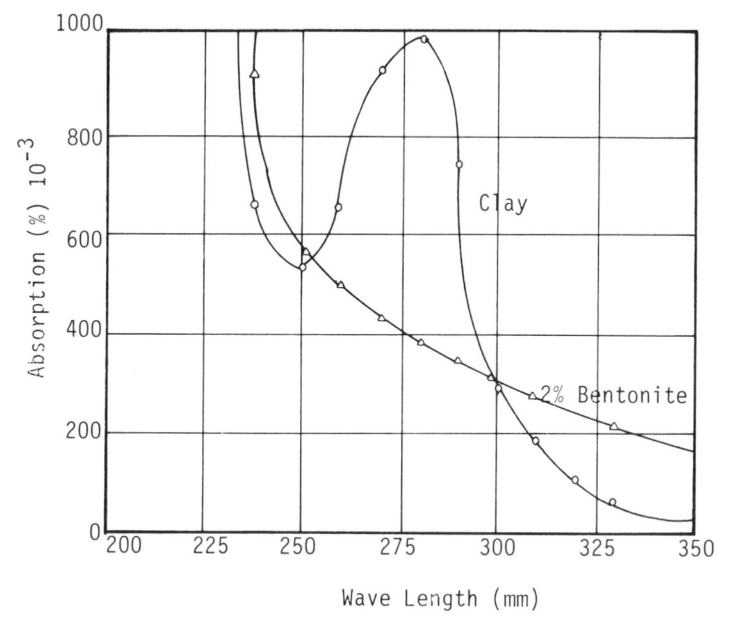

Figure 10. Chemical Absorption by Soil Medium

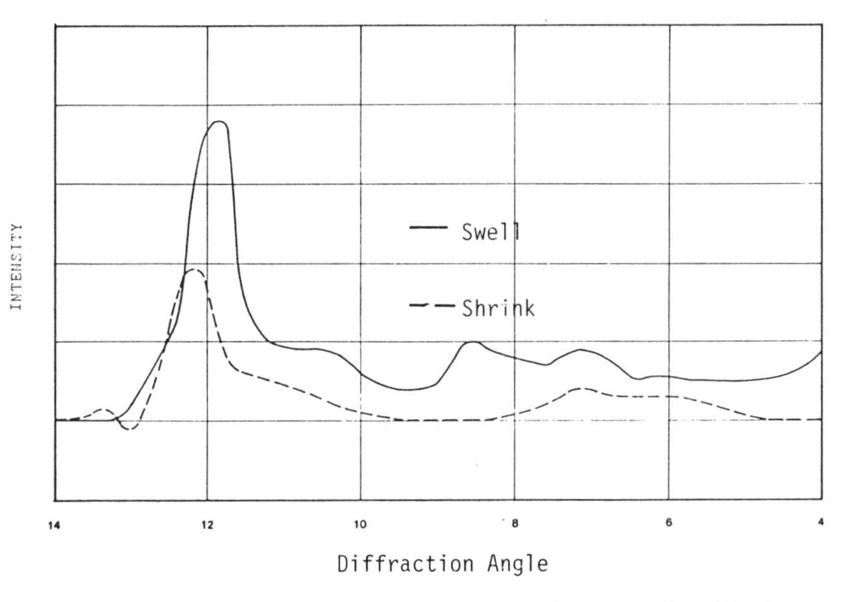

Figure 11. X-ray Diffraction Pattern - 2% Bentonite Blend

IMMEDIATE INSITU CLAY LINER PERMEABILITY

Kingsley O. Harrop-Williams,
The BDM Corporation, McLean, Virginia

INTRODUCTION

Of the many liner materials presently in use to contain hazardous wastes, the local availability of suitable clay soils in many regions of the United States makes the use of soil liners an economically attractive design alternative. In addition, clay liners will always be used as backup for synthetics. The assurance that a compacted liner satisfies an acceptance criteria is related to itş permeability and real-time monitoring of the construction of a clay liner requires immediate feedback on its permeability. Unfortunately, it is very difficult and time consuming to conduct permeability testing either insitu or on insitu samples taken to the laboratory. Therefore, for real-time control of clay placement to achieve a desired permeability, other parameters must be measured and the permeability implied.

The most frequently used technique for the control of earth placement is to monitor the placement moisture content and dry unit weight, and correlate them to the desired property. This is typically done for strength and displacement characteristics, and to a lesser extent for permeability. The obvious advantage for such an approach is direct monitoring. Direct monitoring can be performed as part of the compaction process. This provides all concerned parties (designer, builder, regulatory, and owner) with nearly immediate feedback (in comparison to permeability testing) regarding the acceptability of the liner construction. If a criterion relating moisture content and dry unit weight to permeability is valid, then all that is necessary is for the clay liner to be compacted to the design dry unit weight and moisture content.

Geotechnical and Geohydrological Aspects of Waste Management, D. J. A. van Zyl et al., Eds., © 1987 Lewis Publishers, Inc., Chelsea, Michigan – Printed in USA.

In this paper, a theoretical relationship is developed between the permeability and the easily measured dry unit weight and moisture content of the clay liner. The development assumes that due to the low permeability of compacted clays, the soil never attains complete saturation and the line of optimum is below the zero air voids curve. The analysis then follows from the view point of flow through unsaturated porous media.

ROLE OF CAPILLARY PRESSURE

For convenience the pore water in the soil is thought of as travelling through a tortuous narrow channel in which capillary pressure develops. By considering the equilibrium of forces in this channel, it can be shown that the capillary pressure can be written as (1)

$$Pc = 2\sigma \cos\beta / R \tag{1}$$

where R is the radius of the channel, σ is the surface tension, and β is the angle between the extended fluid meniscus and the walls of the channel.

In unsaturated flow the radius R is replaced by the hydraulic radius

$$Rh = R/2 \tag{2}$$

The square of the hydraulic radius is proportional to the wetted area, which in turn is proportional to the degree of saturation, S, of the channel. Hence equation (1) can be written as

$$Pc = \sigma \cos\beta / (S/C)^{1/2} \tag{3}$$

where C is a proportionality constant. Upon differentiating this equation and dividing by the capillary pressure, we find

$$dPc/Pc = -dS/(2S) \tag{4}$$

The integration of this equation, and imposing the boundary condition Pc = Pb at S = 1, gives the degree of saturation as

$$S = (Pb/Pc)^{\lambda} \tag{5}$$

where $\lambda = 2$. This equation was originally discovered experimentally by Brooks and Corey (2) who called Pb the bubbling pressure (air pressure needed to force air through an initially water saturated sample).They found λ =2 for typical porous media, λ less than 2 for soils with well developed structures, and λ greater than 2 for sands. This parameter was referred to as a pore size distribution index.

For the saturated condition Pb = Pc using the full radius of the channel, or from equation (1) it is

$$Pb = 2\sigma \cos\beta / R \quad (6)$$

Hence, substituting equation (6) into (5) then into (1) the hydraulic radius is obtained as

$$Rh = 0.5RS^{1/\lambda} \quad (7)$$

ROLE OF THE DEGREE OF SATURATION

The seepage velocity of flow in a tube can be derived theoreticaly as (8)

$$v_a = k_a i \quad (8)$$

where

$$k_a = Rh^2 \gamma_f / (Cs\, \mu\, T^2) \quad (9)$$

and i is the hydraulic gradient. Here γ_f is the unit weight of the fluid, μ is its viscosity, T is the toruosity of the channel (defined as the ratio of the path length between two pionts to the distance between the points), and Cs is a shape factor that depends on the cross-sectional shape. For circular cross-sections Cs = 2, and for square ones Cs = 1.78.

The seepage velocity applies to the wetted area of the tube. The general velocity on which permeability is based is

$$v = v_a A_f / A_t \quad (10)$$

where A_t is the cross-sectional area of the channel, and A_f is its wetted cross-sectional area. The ratio A_f / A_t is equal to the ratio of the liquid volume to the soil volume. This in turn is equal to the product of the degree of saturation and the porosity,n; or

$$A_f / A_t = Sn \quad (11)$$

Substituting equation (9) and (11) into (10) gives

$$v = ki \quad (12)$$

where the permeability is

$$k = Rh^2 \gamma_f\, nS / (Cs\, \mu\, T^2) \quad (13)$$

Burdine (3) showed experimentally, and Wyllie and Gardner (9) showed theoretically that the tortuosity at saturation S can be written in terms of that at complete saturation as

$$T = T_1/S \tag{14}$$

where T_1 is the tortousity at complete saturation. Wyllie and Spangler (10) found that T_1 is about 1.41 for typical granular porous media. The influence of degree of saturation on the permeability can now be evaluated by substituting equations (7) and (14) into equation (13); that is

$$k = Bn(\gamma_f/\mu)S^{\alpha} \tag{15}$$

where

$$\alpha = (3\lambda + 2)/\lambda \tag{16}$$

and

$$B = R^2/(4CsT_1^2) \tag{17}$$

are grain size distribution and channel geometry constants, respectively.

ROLE OF COMPACTION

The degree of saturation of a soil can be expressed in terms of the moisture content w, the specific gravity of the solids G, the unit weight of water γ_w, and the dry unit weight of the soil γ_d as (1)

$$S = \gamma_d wG/(G\gamma_w - \gamma_d) \tag{18}$$

Also one can express the porosity as

$$n = 1 - \gamma_d/(G\gamma_w) \tag{19}$$

Hence, in terms of the compaction variables γ_d, and w, the permeability becomes

$$k = B(\gamma_f/\mu)[1-\gamma_d/(G\gamma_w)][\gamma_d wG/(G\gamma_w - \gamma_d)]^{\alpha} \tag{20}$$

COMPARISON WITH EXPERIMENTAL DATA

Experimental comparison of the general form of the equation was made using the data provided by Lambe (5) on compacted Siburua Clay as shown in figure 1. First it was recognized that due to the small range of dry

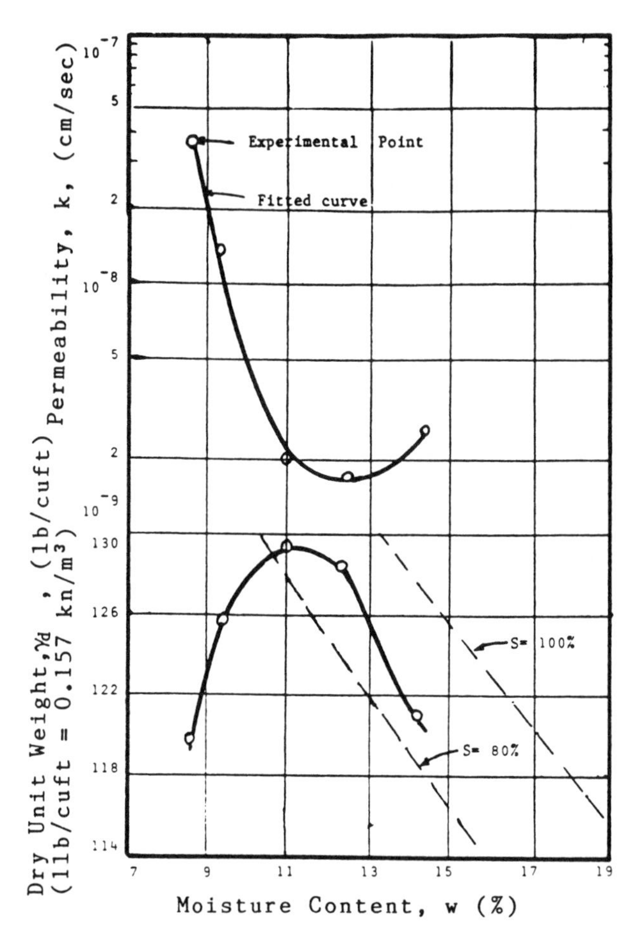

Figure 1. Fitted Permeability Curve (Data from 5)

unit weights experienced during compaction the porosity was approximately constant. Hence equation (20) can be expressed in the linear form

$$\ln k = A + \alpha \ln S \tag{21}$$

where S is as given in equation (18), and A and α can be determined as regression constants. A regression of this equation and the data presented in figure 1, assuming G = 2.7 (a recommended value for clay) gave A = -19.873 and α= -5.19 with a coefficient of correlation of 99.04%. For comparison, the proposed relationship (equation 20) is plotted with the experimental points in figure 1.

DISCUSSION

In equation (20) the primary parameters dependent on the fluid are its viscosity and unit weight. This is true for steady state flow only. Before steady state is reached some fluids cause chemical exchanges with the soil that may affect the structure (shape and tortuosity of the pore channels) of the clay. The structure of the clay is one of the most influential parameters to its permeability. This was highlighted by Lambe and Whitman (6) and mitchell et al. (7). It depends also on the method and effort of compaction (4).

The developed relationship between permeability and the easily measured dry unit weight and moisture content allows clay liners to be tested as they are being built. The empirical structure-specific parameters A and α, being sensitive to compaction effort and method of compaction, should be obtained by regression beforehand using laboratory determined permeability values of field compacted soils. With equation (21) and a fast moisture content and dry unit weight measuring device like the nuclear densometer, Immeniate permeability values can be obtained throughout the liner.Hence direct monitoring of construction is possible and increased sampling facilitated.

REFERENCES

1. Bowles, J. E., Physical and Geotechnical Properties of Soil (New York: McGraw-Hill Book Co., 1984).

2. Brooks, R. H., and A. T. Corey. "Properties Porous Media Affecting Fluid Flow," ASCE J. of Irrig. & Drain., Vol. 92, IR2, (1966).

3. Burdine, N. T., "Relative Permeability Calculations from Pore Size Distribution Data," Trans. AIME, Vol. 198, (1952).

4. Harrop-Williams, K. O., "Clay liner Permeability Evaluation and Variation," ASCE J. of Geo. Eng., Vol. 111, No. 10, pp 1211-1225, (1985).

5. Lambe, T. W., "The Engineering Behavior of Compacted Clay", ASCE J. of Soil Mech. & Fdn. Eng., Vol. 84, SM2, pp 1655-1, (1958).

6. Lambe, T. W., and R. V. Whitman, Soil Mechanics (New York: John Wiley & Sons, Inc., 1969).

7. Mitchell, J. K., D. R. Hooper, and R. G. Campenella, "Permeability of Compacted Clay," ASCE J. of Soil Mech. & Fdn. Eng., Vol. 91, SM4, pp 41, (1965).

8. Wu, T. H., Soil Mechanics, (Boston: Allyn & Bacon, Inc., (1976).

9. Wyllie, M. R., and G. H. Gardner, "The Generalized Kozeny-Carman Equation, A Novel Approach to Problems of Fluid Flow," World Oil Prod., Sect. 210-228, (1958).

10. Wyllie M. R., and M. B. Spangler, "Applications of Electrical Resistivity Measurements to Problems of Fluid Flow in Porous Media," Bull. Amer. Assoc. of Petrol. Geol., Vol.36, (1952).

PERMEABILITY TESTS FOR HAZARDOUS WASTE MANAGEMENT UNIT CLAY LINERS

Hsien W. Chen and Leonard O. Yamamoto
International Technology Corporation, Irvine, California

ABSTRACT

Field and laboratory permeability tests were conducted on a 36 m x 24 m x 0.76 m clay fill designed and built using the same materials and construction methods employed for a hazardous waste landfill clay liner. The test fill was built of a fat clay (CH). The BAT porous probes and the sealed double-ring infiltrometer (SDRI) were used to perform the field permeability tests. Laboratory permeability tests using a triaxial test cell were performed on shelby-tube samples from the liner to study the effect of applied effective confining stresses on the measured coefficients of permeability.

Results of the testing indicated coefficients of permeability for both the SDRI and the BAT probe tests to be approximately one order of magnitude higher than the laboratory-determined values. It is suggested that this difference is primarily due to the unrestrained swelling of the clay which occurred during the field testing. It is estimated the swelling caused a decrease in the dry density of the clay liner of 20 percent. It is concluded that laboratory permeability tests, with the ability to apply consolidation stresses simulating the design waste fill overburden, may be more suitable than field tests in evaluating the performance of well-constructed, high-swelling clay liners.

INTRODUCTION

The liner systems of hazardous waste management units are normally constructed of both earthen and synthetic liners. While there are standard, generally accepted field test methods to evaluate the construction quality of synthetic liners, this is

Geotechnical and Geohydrological Aspects of Waste Management, D. J. A. van Zyl et al., Eds., © 1987 Lewis Publishers, Inc., Chelsea, Michigan—Printed in USA.

not so for earthen liners. Compacted clay liners are principally evaluated by their permeability (hydraulic conductivity). Testing of clay liners has generally relied upon laboratory permeability tests. Prompted by regulatory requirements and by the recognition that laboratory permeability tests may result in much lower values than field tests [1,2], there has been an effort to place a heavier reliance on field permeability testing.

A clay test fill (liner) was constructed in conjunction with a hazardous waste landfill for the purpose of performing field permeability tests. Two recently developed field permeability test methods, the sealed double-ring infiltrometer (SDRI) and the BAT porous probes, were used to evaluate the test liner. Laboratory permeability tests using a triaxial test cell were performed on relatively undisturbed Shelby-tube samples from the test liner. The results of the laboratory testing were compared to the field test results and also were used to study the effect of changing the applied effective confining stress on the measured coefficients of permeability.

This paper describes the construction methods and engineering properties of the clay test fill and presents a description of each of the permeability tests performed. Each of the test methods and their results are compared and discussed. Limitations of each system are presented along with specific recommendations for future testing.

CLAY TEST FILL

The clay test fill was constructed using the same clay materials, equipment, and procedures used in the construction of an actual landfill clay liner. The completed clay test fill was 36 m long, 24 m wide, and 0.76 m thick. In-place density tests were performed during construction to monitor the moisture and density condition of the test fill. Bulk samples were obtained from each construction lift for laboratory tests. The soil properties and the construction method of the test fill are described below.

Soil Properties

As part of the construction quality control, a comprehensive laboratory testing program was implemented to determine engineering properties of the soil used in the test fill. Test results are presented in Table 1. In addition, seven x-ray diffraction tests were performed to provide a description for the clay mineralogy. The test results indicated the soils used in the test fill contained approximately 10 to 50 percent clay, with smectite (montomorillonite) composing 25 to 50 percent and kaolinite 10 to 25 percent of the clay fraction.

Table 1. Summary of the Clay Material Laboratory Test Results

Soil Property Description	Test Designation	Number of Tests	Test Results
Liquid Limit	ASTM D 4318	6	53 to 63%
Plasticity Index	ASTM D 4318	6	28 to 41%
Grain Size	ASTM D 422	6	92 to 100% <0.075 mm 52 to 59% <0.002 mm
Specific Gravity	ASTM D 854	5	2.68 to 2.74
Maximum Dry Density	ASTM D 1557	5	18.1 to 18.6 kN/m^3
Optimum Moisture Content	ASTM D 1557	5	14 to 15%
Expansion Index	UBC Standard No. 29-2	3	136 to 156

The soil, in its native state, was described as very stiff, reddish brown, moist, inorganic, highly expansive, fat clay with U.S.C.S. classification of CH.

Method of Construction

The clay material was excavated from an on-site borrow pit near the test fill. After spreading, the material was disced to break down the size of clods to allow more uniform compaction and to hasten drying. The clay moisture content was measured using a microwave oven which was calibrated against a conventional laboratory oven in accordance with ASTM D 2216.

The 0.76-m thick test fill was constructed in five lifts of approximately equal thickness. A level was used to control the thickness of each lift. A 0.15-m thick clean sand covered with a geotextile was constructed below the test fill. Although it is not a constituent of the actual clay liner in the landfill, the sand layer was designed to define the lower boundary condition for the SDRI test. The clay fill was compacted by a CAT 815, a self-propelled tamping foot compactor. The area of coverage and number of passes made by the compactor were observed and recorded.

A total of 46 field moisture and density tests were performed. The density tests included 30 tests by nuclear method (ASTM D 2922), eight verification tests by rubber-balloon method (ASTM D 2167) and eight verification tests by sand-cone method (ASTM D 1556). Of the 46 density tests performed, none of the results fell below the specified 90 percent of maximum dry density. The results also indicated that the test fill was constructed with moisture contents approximately 1 to 5 percent wet of optimum. The completed test fill was covered with a plastic sheet to prevent desiccation.

LABORATORY PERMEABILITY TESTING

Laboratory permeability (hydraulic conductivity) tests were conducted on shelby-tube samples from the test fill. The test method used was the EPA Method 9100 "Triaxial-Cell Method with Back Pressure" [3]. The test procedure is well documented in the literature and is briefly described below.

Test Procedure

Each specimen, 73 mm in diameter, was extruded from the sample tube, trimmed to about 76 mm in length, surrounded on the sides with a double flexible membrane, and sealed with rubber O-rings. With an effective confining pressure of 21 kPa, the sample was subjected to a percolation water head of 48 kPa at the bottom in an attempt to purge air from the soil. The percolation process continued for three weeks until water flowed out from the sample top. A 345 kPa back-pressure was then applied on both ends to saturate the sample. The back pressure was increased in 35 kPa increments. A minimum of 8 hours was allowed between increments to avoid sample overconsolidation. Local well water, the same as that used for the SDRI test, was used in the laboratory tests.

A hydraulic gradient of approximately 180 was used to generate flow in a reasonable time. The gradient resulted from a pressure difference applied across the sample length. Therefore, the sample was nonuniformly consolidated with the effective confining stress higher at the top than at the bottom end. An effective stress of 172 kPa was applied to the sample top (outflow) end to simulate a design waste overburden of 9 m. The amount of discharge was collected and measured in a pressurized and enclosed burette system outside the test cell. Discharge was collected and measured for about 40 days to assure steady-state flow before the test was terminated. Four clay samples were tested in accordance with this procedure.

Another three samples were tested using the same procedure except that each test was performed in four separate stages. At each stage, the effective confining stress was increased to determine the effect on the measured coefficients of permeability. The four effective stresses at the outflow end were 55, 104, 207, and 345 kPa. The initial percolation pressure head for these tests was 35 kPa. The calculated hydraulic gradient varied from 20 to 75 for the samples at different stages of the testing.

At each stage, sufficient time was allowed for consolidation prior to measuring the discharge for permeability calculations. Each of these tests lasted 34 days before termination.

Test Results

The calculated permeability coefficients and dry densities for the clay samples used in the four single-stage tests are presented in Table 2. It was noted that swelling occurred to each of the four samples, causing a decrease of approximately 6 percent in the dry densities.

Results of the three multiple-stage tests are shown in Figure 1. The calculated permeability coefficients for the samples at an effective stress of 55 kPa are approximately one order of magnitude higher than those calculated for a stress of 345 kPa. The difference is significantly higher than that reported by others [4]. It is likely the result of a decreased void ratio and increased tortuosity in the high-swelling clay under higher stresses.

SEALED DOUBLE-RING INFILTROMETER (SDRI)

The SDRI has recently evolved from the double-ring infiltrometer (ASTM D 3385). The major improvement is a sealed, dome-shaped inner ring which can be submerged within the ponded water of the outer ring to reduce the effects of evaporation and temperature fluctuations. Only limited experiences of SDRI applications have been reported in the literature [5].

Table 2. Calculated Density and Permeability Coefficients for the Single-Stage Tests (with Effective Confining Stress of 172 kPa).

Sample Number	Permeability Coefficient at 20°C (cm/s)	Sample Dry Density Before Test (kN/m^3)	Sample Dry Density After Test* (kN/m^3)
TP-1	1.0×10^{-9}	17.7	16.9
TP-2	1.3×10^{-9}	17.3	16.5
TP-3	2.0×10^{-9}	17.2	16.2
TP-4	1.5×10^{-9}	17.6	16.5

* Sample dimension measured immediately after termination of tests.

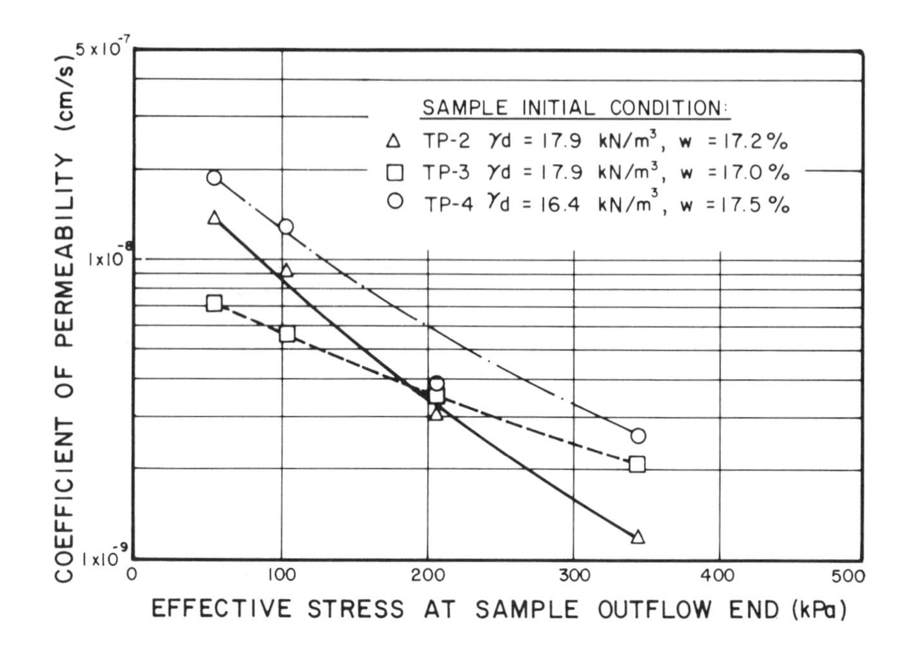

Figure 1. Coefficient of Permeability versus Effective Stress at Sample Outflow End.

Method of Installation

The specific SDRI used in this test consisted of two fiberglass rings with diameters of five and seven feet. Mariotte tubes similar to that described in ASTM D 3385 were used for most of the test to measure flow as well as maintain constant and equal water level within the inner and outer rings. A separate Mariotte tube was used for each ring. A flexible intravenous bag connected to the inner ring was used for the same purposes during the last 11 days of the test.

Both rings were embedded in two-inch wide trenches. The inner and the outer trenches were excavated to a depth of three and six inches, respectively. To seal the rings within the trenches, the trenches were filled with a soil/bentonite mixture having a permeability coefficient of 2×10^{-8} cm/sec or less.

Local well water was used to fill the rings for the test. To reduce evaporation and temperature variations, the entire water surface was covered with styrofoam and plastic sheeting. Three digital temperature probes with 0.1°C resolution were used in this test to monitor water temperature in the rings. Six BAT probes were installed at 6 to 12 inches below the clay fill surface within the annular space of the two rings. The probes were used to monitor the capillary potential in the unsaturated clay to allow the calculation of hydraulic gradient. The installation of the SDRI and BAT probes is shown in Figures 2 and 3.

Figure 2. BAT Probe Risers, the Dome-Shaped Inner Ring of the SDRI and the Excavated Trench for the Outer Ring.

Figure 3. Completed SDRI Installation with Styrofoam Insulation.

Infiltration Rate

The infiltration rate was calculated by:

$$I = \frac{Q}{tA}$$

where

I = infiltration rate (cm/s)
Q = measured flow (cm^3)
t = interval of time corresponding to the measured flow Q, (s)
A = area of infiltration (cm^2).

A plot of the calculated infiltration rate as a function of time for the inner ring is shown in Figure 4.

Field observations indicated that unresponsiveness and inconsistency of the Mariotte system caused variable drops in the water level in the outer ring. Such a drop in the outer-ring water level causes a differential pressure between the two rings and results in an increased flow from the Mariotte system to the inner ring due to expansion of the inner ring and/or leakage from the inner to the outer ring. It is believed that the wide scatter of the infiltration rate was caused primarily by this observed phenomenon. Volume and viscosity changes of the water caused by temperature variations may have been among the other factors contributing to the test error. Because of the problem associated with the Mariotte system, it was abandoned for each of the rings. A water-filled flexible bag was used in place of the Mariotte system to measure flow into the inner ring. The amount of flow was determined by measuring the weight difference of the water-filled bag over a known time interval. The outer ring was manually filled to maintain a constant water depth of 46 cm.

The inner ring infiltration rate became relatively steady after the flexible bag was put into use; however, the outer ring manual-filling rate was noticeably increasing with time. Eventually wet marks appeared outside and around part of the outer ring. Exploratory pits revealed that water was seeping laterally from the outer ring at a depth of six inches or more below the soil surface. This matched the depth of embedment of the outer ring. The test was then terminated 44 days after the rings were first impounded.

A post-test investigation was conducted. Its major findings are listed below:

- The ground surface in the rings swelled after inundation. It became approximately 7 cm higher than the surrounding unwetted area.

- Penetration with a sharp-edged steel probe indicated a relatively uniform wetting front had reached about 36 cm below the original test fill surface or 43 cm below the swelled surface.

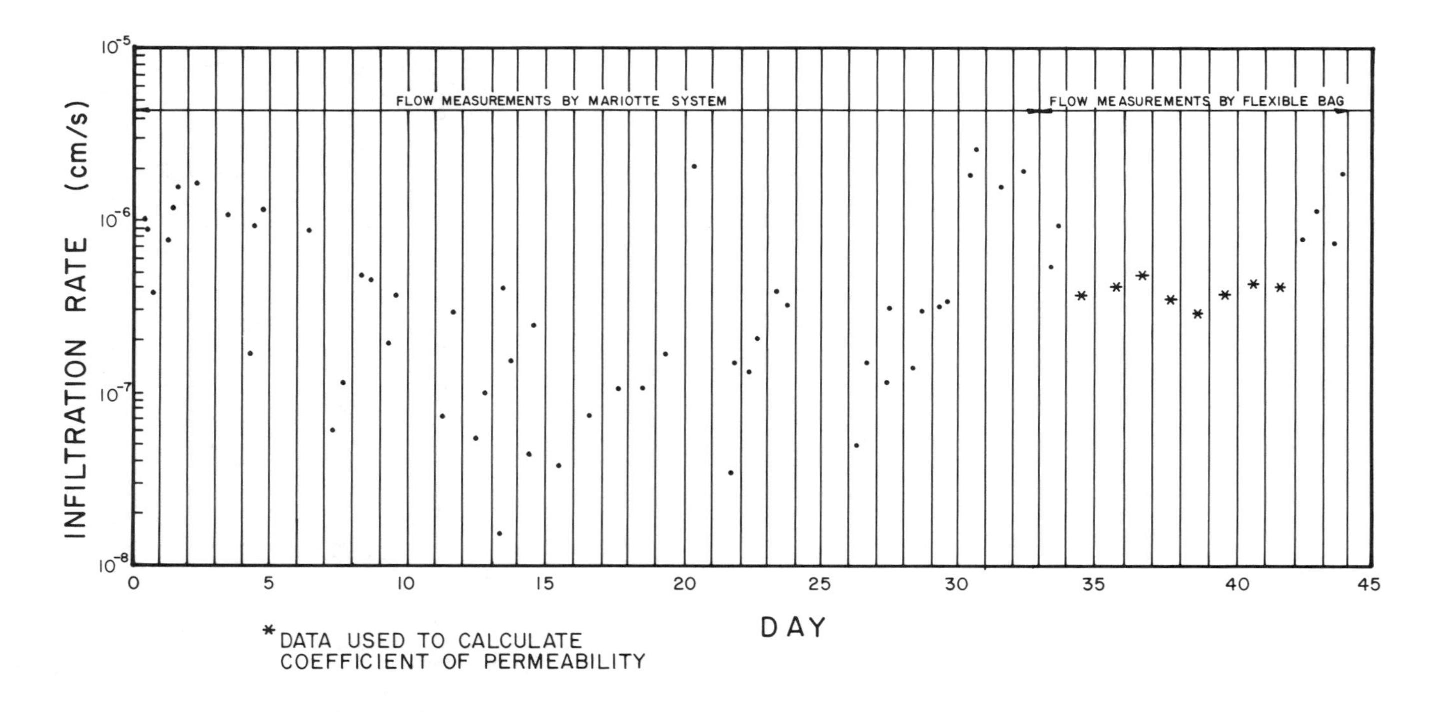

Figure 4. The SDRI Test Inner Ring Infiltration Rate

- Shelby-tube and bulk soil samples were collected from the inner ring area. Laboratory test results of these samples showed that at 0 to 20 cm below the swelled ground surface, average moisture content and dry density were 35 percent and 13.3 kN/m^3, respectively. This was more than a 20 percent decrease of dry density of the clay test fill. The moisture content decreased and the density increased with increasing depth. At a depth of 38 to 51 cm below the swelled surface, the clay remained in its pre-test condition.

- Leaks were detected around the port where the temperature probe wires lead into the inner dome (ring) and around two other ports. These leakage pathways were likely responsible for most, if not all, of the higher inner ring infiltration rates measured at a time when the outer ring water level was low.

Coefficient of Permeability

To convert the infiltration rate to a coefficient of permeability, the hydraulic gradient and lateral spreading factor should be known. Thus, the coefficient of permeability is calculated by:

$$K = \frac{I}{i} F$$

where

K = coefficient of permeability (cm/s)
I = infiltration rate (cm/s)
i = hydraulic gradient (cm/cm)
F = lateral spreading factor.

The infiltration rate is obtained by direct measurement as described previously. The estimation of hydraulic gradient and lateral spreading factor are described below.

The hydraulic gradient may be defined as the ratio of the total water head loss to the travel distance of the wetting front. Total water head consists of elevation (gravitational) head and pressure head. In partially saturated soil, capillary head (negative pressure) also contributes to the total head available to drive flow. The six BAT probes installed in the annular space measured an average capillary head of 6.2 m of water prior to the start of the SDRI test. Since the wetting front was far from reaching the dry sand at the bottom of the test fill, it would be unreasonably conservative to assume zero capillary potential for the hydraulic gradient calculation. By assuming a capillary potential of 6.2 m immediately below the idealized wetting front, the hydraulic gradient is calculated by:

$$i = \frac{h + d + h_s}{d} = \frac{46 + 43 + 620}{43} = 16.5$$

where

i = hydraulic gradient (cm/cm)
h = depth of ponded water above the swelled surface of fill (cm)
d = thickness of clay above the wetting front (cm)
h_s = suction below the wetting front (cm)

According to the SDRI developer [5], the lateral spreading factor F is 0.9 for this test set-up. However, the recommended F factor was derived by assuming saturated, steady-state flow throughout the entire liner thickness and zero capillary head at the bottom of the liner [5,6]. In our opinion, this and other associated assumptions, do not apply to this test fill or to clay liners in general. Therefore, the F factor is not included in this analysis.

The average infiltration rate calculated from flexible-bag readings is 4.5×10^{-7} cm/sec. The last four flexible-bag data points were excluded owing to their association with very high lateral flow which ended the test. Excluding the F factor, the permeability coefficient K for the SDRI test is calculated as:

$$K = I/i = (4.5 \times 10^{-7} \text{ cm/sec})/16.5 = 2.7 \times 10^{-8} \text{ cm/s}$$

BAT POROUS PROBES

The BAT system used in this project consists of a plastic tip containing a cylindrical porous filter driven into the test fill at the end of an extension pipe. The filter is connected to a container partly filled with water and partly filled with gas. The rate of flow of liquid through the porous filter is computed by measuring the gas pressure change in the container and then applying Boyle's law. Analysis of the time/pressure record utilizing the falling head method yields the coefficient of permeability. Capillary potential of the unsaturated clay was included in the analysis. Details regarding the description, theory and applications of the BAT system are reported elsewhere [7].

Six BAT probes were installed prior to the installation of the SDRI. The tips were driven into the test fill using a slide hammer. They were located in the area of the annulus where the infiltrometer rings would be located. Three probes were set at a depth of 15 cm and the other three were set at 30 cm. The riser pipes shown in Figures 2 and 3 indicate locations of the probes.

Two sets of field permeability tests were conducted with the same set of porous probes. The first set of tests were performed before water was impounded for the SDRI test. At the time of the second testing, the wetting front of the SDRI probably had passed all but one of the 30-cm deep probes. The test results are presented in Table 3.

As shown in Table 3, the first set of coefficients of permeability appear to be similar to those measured by laboratory testing of shelby-tube samples shown in Table 2. Permeability coeffiients calculated from the second set of tests appear to be one to two orders of magnitude higher than the first set for each probe.

The second set of data may reflect the lower clay densities resulting from unrestrained swelling and leakage, if any, around the probes due to disturbance of the probes during the SDRI test.

A post-test investigation was conducted to examine the clay seals around the BAT probes and the probes themselves. During the careful excavation, no apparent cracks were observed in the clay around the 30-cm deep probes. The condition of the seals around the 15-cm deep probes could not be determined because of the great disturbance caused by the removal of the outer ring.

Table 3. Summary of BAT Field Permeability Test Results.

Probe	Test No.	Depth (cm)	Pore Pressure Before Test (m H_2O)	Coefficient of Permeability (cm/s)
IV-6	1	15	-6.80	8.4×10^{-10}
	2	15	+0.18	1.9×10^{-7}
IV-7	1	30	-6.46	2.0×10^{-9}
	2	30	+0.58	2.2×10^{-8}
IV-8	1	15	-4.51	1.7×10^{-9}
	2	15	+0.79	6.9×10^{-8}
IV-9	1	30	-8.11	6.2×10^{-10}
	2	30	--	No Test
IV-10	1	15	-6.80	9.3×10^{-9}
	2	15	+0.55	7.2×10^{-9}
IV-11	1	30	-4.63	8.4×10^{-10}
	2	30	-1.19	2.8×10^{-8}

DISCUSSION

Each of the three methods adopted in this investigation has its advantages and limitations for application on hazardous waste clay liners. The theoretical bases, testing equipment and testing techniques of each method are discussed here.

Laboratory Permeability Testing

The laboratory permeability test using a triaxial test cell is probably the most widely-used and scrutinized method for testing clay samples. It has the ability of saturating the sample with back pressure, simulating in-situ overburden pressure, and generating measurable amounts of outflow by applying large hydraulic gradients. The testing procedures and techniques are well documented in the literature. Variables and their effect on measured permeability have been studied [1,4]. The variables studied have included smear zones on sample ends, sample length to diameter ratio, gradient magnitude, effective confining stress, growth of microorganisms, etc. The major criticisms regarding its application to hazardous waste clay liner have been the small sample size tested and the usually lower permeability measured compared to field tests. It is understandable that laboratory-measured permeability is much lower in value than the permeability calculated from the loss of water ponded over poorly constructed and/or poorly maintained clay liners. This is likely due to large desiccation cracks and holes resulting from decayed organics.

Sealed Double-Ring Infiltrometer

Recently evolved from the regular double-ring infiltrometer with the improvements to minimize evaporation and reduce temperature fluctuations, the SDRI shows promise in measuring infiltration rate for materials with low permeability. The SDRI method is preferred by regulatory agencies for two reasons: (1) it is a field test and (2) its sample size is large as compared to laboratory test samples. However, experience gained from this study indicates that further refinements to the apparatus are needed to increase the consistency of the measuring device and to maintain a constant water level in both rings to reduce the error caused by differential pressures.

The most prominent question regarding the SDRI method is a generic one common to all infiltrometer tests; how does one properly interpret the test data and convert the infiltration rate of a partially saturated, transient flow regime into a saturated hydraulic conductivity? The analysis used in this investigation is a simplified one. However, it incorporates the thickness of the wetted soil profile and the capillary potential existing in the partially saturated clay below an idealized wetting front.

In the process of calculating the coefficient of permeability from an infiltration test result, an implicit assumption is made that infiltration is equal to discharge for a defined section of the wetted soil column. This assumption may involve a substantial amount of error for swelling clays due to the reason that part of the infiltration will be held by the lower portion of the soil column to fill the increased pore volume that is created by the swelling process.

The lateral spreading factor F, as previously discussed, is another subject that needs to be studied. A two-dimensional mathematical modeling of a partially saturated flow regime with the verification by laboratory model tests would be a realistic approach.

It is our opinion that further research is needed to refine the theoretical bases for the SDRI testing method in order to reduce the uncertainties involved in the test data reduction process.

BAT Porous Probes

The BAT method basically measures the horizontal permeability by generating radial flow from the porous ceramic probes. It incorporates the measured capillary potential and size of the wetted bulb in the data reduction process. Attention should be given to the probe installation procedure to minimize soil disturbance while obtaining a tight seal around the probes. Periodic calibration of the electronic data acquisition system in essential. It is considered a "point" test because the volume of soil involved in each test is small. However, statistical analysis may be conducted to estimate the number of tests deemed to be sufficient for providing information to characterize the soil. Similar to the other field test methods, the design overburden stress acting upon the clay liner cannot be simulated by the BAT testing method.

SUMMARY, CONCLUSIONS, AND RECOMMENDATIONS

The average value of coefficients of permeability obtained from seven laboratory tests using a triaxial test cell with an effective stress simulating the design waste overburden was approximately 3×10^{-9} cm/s. The value is similar to the ones measured using BAT probes on the as-placed clay test fill. The second set of BAT tests, conducted after the clay column was wetted by the ponding water of the SDRI test, reported an average permeability of 3×10^{-8} cm/s, excluding one exceptionally high value. This average value is similar to the 2.7×10^{-8} cm/s reported by the SDRI test and it is approximately one order of magnitude higher than the laboratory test results. By examining these and the multiple confining stress test results, it is suggested that the difference in permeability is primarily due to the unrestrained swelling of the clay which occurred during the field testing.

It is concluded that simulating design stress conditions is essential for both the laboratory and field permeability tests. Laboratory tests performed with proper consolidation stress may be more suitable than field tests in evaluating the performance of well-constructed, high-swelling clay liners.

During soil liner compatibility testing with waste leachate, it is a general practice to apply a high hydraulic gradient, and therefore, an unavoidably high consolidation stress, to accelerate the test and achieve two pore volumes of accumulated outflow. Realizing the effect of consolidation stress on the measured permeability of high-swelling clays, it is recommended that a low hydraulic gradient be used for the permeability testing. This will require that permeability tests be run separate from compatibility tests.

ACKNOWLEDGMENTS: The writers wish to thank Edward Castellanos and Jagdish Mathur for reviewing the manuscript and Thierry Sanglerat of Earth Technology Corporation for his suggestions regarding the multiple-stage laboratory permeability tests.

REFERENCES

1. Olson, R.E., and D.E. Daniel, "Measurement of the Hydraulic Conductivity of Fine-Grained Soils," ASTM STP 746, 1981, pp. 18-64.

2. Daniel, D.E., "Predicting Hydraulic Conductivity of Clay Liners," Journal of Geotechnical Engineering Division, ASCE, Vol. 110, No. 2, February, 1984, pp. 285-300.

3. "Proposed Sampling and Analytical Methodologies," U.S., EPA Report SW-846, 2nd Edition, 1984.

4. Carpenter, G.W., and R.W. Stephenson, "Permeability Testing in the Triaxial Cell," ASTM Geotechnical Testing Journal, Vol. 9, No. 1, March 1986, pp. 3-9.

5. Daniel, D.E., and S.J., Trautwein, "Field Permeability Test for Earthen Liners," Proceedings of In Situ '86, ASCE specialty conference, Blacksburg, Virginia, June 23-25, 1986, pp. 146-160.

6. Day, S.R., and D.E. Daniel, "Field Permeability Test for Clay Liners," ASTM STP 874, 1985, pp. 276-288.

7. Petsonk, A.M., "The BAT Method for In Situ Measurement of Hydraulic Conductivity in Saturated Soils," M.S. Thesis in Hydrogeology, University of Uppsala, Sweden, 1984.

LABORATORY TESTING OF BENTONITE AMENDED SOIL MIXTURES PROPOSED FOR A MINE WASTE DISPOSAL FACILITY LINER

John F. Wallace, P.E.
Dames & Moore, Salt Lake City, Utah

INTRODUCTION

Little data exists in the literature concerning the hydraulic conductivity (permeability) of bentonite amended soil mixtures [1-4]. No laboratory data exists exploring the effects of either cyclic freeze/ thaw or desiccation on the performance of such potential liner materials. This paper presents the results of a comprehensive laboratory testing program directed at developing a design basis for a bentonite amended soil liner proposed for a mine waste disposal facility. The test program was developed to replicate anticipated field construction details with specific attention placed on sequence of liner material processing, the use of site water and the pore fluids brought into contact with test specimens during testing.

TEST PROGRAM DEVELOPMENT

The laboratory testing program focused on the need to develop a consistant liner material having a coefficient of hydraulic conductivity or permeability of $5x10^{-8}$ cm/sec using on-site soils. A phased test program approach was used to permit a systematic refinement of the design basis.

Geotechnical and Geohydrological Aspects of Waste Management, D. J. A. van Zyl et al., Eds., © 1987 Lewis Publishers, Inc., Chelsea, Michigan – Printed in USA.

Phase I investigated the nature and the spatial variability of near-surface soils across the site. Additionally, a preliminary suite of soil/bentonite mixtures were tested to assess the variation of hydraulic conductivity as a function of bentonite contents ranging from zero to six percent.

Phase II explored the affects of cyclic freeze/thaw and surface desiccation on the hydraulic conductivity of a two percent bentonite amended soil mixture. The affects of increased compaction effort were explored and two long-term tests were initiated.

In Phase III, a composite site fill was prepared from 22 separate bulk samples to represent a potential stockpile source material. From this "source" four subsets of amended soil were prepared to evaluate a refined range of bentonite/soil blends. Test specimens of each subset were prepared at different moisture contents ranging between one and four percent wet of optimum to evaluate hydraulic conductivity sensitivity as a function of molding moisture.

PROPOSED CONSTRUCTION

Proposed facility construction would require excavation at the site to depths ranging 6 to 15 meters. Portions of the excavated materials would be stockpiled off-site for liner construction using a batch technique. A maximum particle size of 19 mm was specified for processed liner materials. The system being considered for liner material processing used a regulating feed conveyor on which dry powdered bentonite would be added to the liner soil prior to entering a mixing box. The bentonite/soil would advance through the mixing box using twin counter rotating screw feed paddle augers. Approximately half way through the mixing, water would be added and mixing continued before discharge into trucks. The liner mix would then be hauled to the site for immediate liner construction. Mixing would take a total of five minutes and material haulage 5 to 10 minutes.

The material preparation procedures discussed in the next section were developed to simulate the processes envisioned for actual construction with specific attention being focused on mixing sequence, mix times and the use of site water.

TEST METHODS AND MATERIAL PREPARATION

General

All index tests including specific gravity, Atterberg limits, and grain-size distribution determinations were performed in accordance with the prescribed ASTM Testing Procedures. Compaction tests were performed in accordance with ASTM D-698, Method C (Standard Compaction Test) or ASTM D-1557, Method C (Modified Compaction Test).

Sample Preparation

Three large bulk samples were prepared for Phase I, each weighting approximately 500 Kg. Prior to processing, representative samples of the prescreened materials were obtained for grain-size testing. Each bulk sample was then screened to remove particles larger than 19 mm (0.75 inches) and returned to storage barrels to maintain field moisture content.

Each screened bulk sample was quartered to form the representative subsamples to be bentonite amended and tested. To each subsample a prescribed amount of bentonite was added and blended at field moisture content for two minutes using a 1/4 cubic yard portable concrete mixer. After mixing was complete, the various mixtures were temporarily stored in plastic bags.

Subsequent moisture conditioning required for compaction testing and test specimen preparation also used the 1/4 cubic yard concrete mixer. A predetermined amount of site water was added to a known dry weight of soil mixture while the mixture was turning. Mixing continued for an additional two minutes subsequent to water addition. Compaction testing or test specimen fabrication proceeded within ten minutes of the completion of mixing without permitting additional time for bentonite hydration.

A composite bulk sample was developed for Phase II from 10 individual test pit bag samples selected to produce a grain-size distribution similar to the narrow range explored in Phase I. Another bulk sample was developed for Phase III study by combining bag samples from 17 test pits selected at random. Further processing, mixing and moisture conditioning of these composite bulk samples was performed in the same manner described for Phase I.

Test Specimen Construction

All permeability test specimens were 100mm in diameter and ranged between 105 and 140mm in height. Each was compacted using a compaction drop hammer to 95 percent of the maximum dry density previously determined for the specific material blend. Molding moisture contents varied but were specifically predetermined for each specimen.

Generalized Permeability Test Procedure

All permeability testing was performed using flexible membrane test cells under constant head conditions. Testing methodology and procedures followed closely with those discussed by Daniel, et al, [5] and Dunn, et al, [6]. Hydraulic gradients applied during testing were typically less than 50. Effective confining pressures were less than 70 kPa. All test specimens were back pressure saturated until achieving a B value of 0.95. Individual tests were performed for a minimum of 48 hours and until a permeability coefficient of within a factor of plus or minus two was achieved for three consecutive two hour readings.

All test specimens were saturated using a synthetic leachate later used as the permeant. The chemical composition of this fluid is presented in Table 1.

Table 1. - Synthetic Leachate Chemistry

Conductivity	3,000	mMhos
Chlorides	14.1	ppm
TDS	2,740	ppm
SO_4	1,800	ppm
Ca	411	ppm
Al	0.20	ppm
Mg	10.2	ppm
Mn	0.01	ppm
pH	4.0	

Desiccation Exposure Procedure

To evaluate the potential deleterious effects of surface desiccation to a compacted bentonite amended soil liner, a series of four test specimens were subjected to varying degrees of atmospheric exposure.

The specimens were encapsulated in latex membranes with only one end exposed for periods of 0, 8, 48, and 96 hours. The average outdoor environmental conditions during the exposure were: temperature-24 to 33 degrees Celcius; wind speed-8 to 32 Km/hr; and visual conditions sunny and clear. At the completion of air exposure, samples were then tested to determine their hydraulic conductivity.

Cyclic Freezing/Thawing

Cyclic freezing and thawing of permeability test specimens was performed in general accordance with the procedures outlined in ASTM D-560 with a specified minimum temperature of -23°C. Each cycle consisted of 24 hours at -23°C followed by 24 hours at 21°C. Each of the test specimens was encapsulated in its own latex membrane and subjected to either 0, 2, 6, or 10 cycles of freeze/thaw prior to placement in triaxial cells for permeability testing.

PHASE I TESTING

A suite of grain-size tests were performed on representative portions of 15 specimens obtained from test pits excavated throughout the 100 acre proposed site. The results of these tests are presented in companion with the grain-size test results for the three large bulk samples A, B, and C prior to screening and further processing in Figure 1.

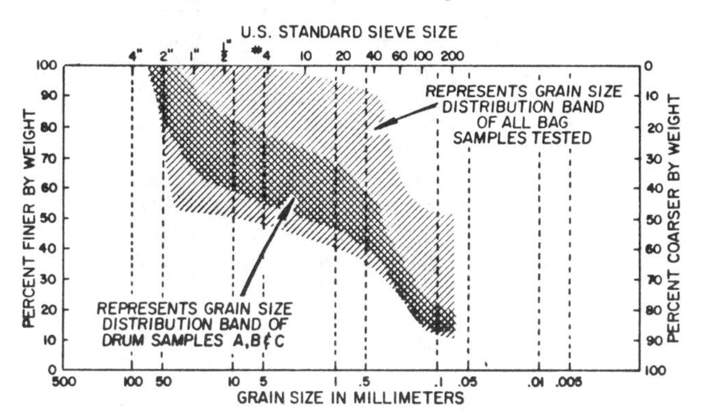

Figure 1. Phase I Grain-Size Analyses Before Material Processing

Index tests including grain-size, specific gravity, Atterberg limits and compaction tests were performed for each of the bentonite amended subsamples prepared. Results of these tests are summarized in Table 2.

Table 2. Summary of Phase I Material Properties

Material Designation	% Bentonite	Grain Size			Compaction* Data		Atterberg Limits			
		-200	D_{50}-mm	D_{10}-mm	γ_dMax	ωopt	LL	PL	PI	G_s
A	0	24.2	0.30	.002	2.12	7.4	16	12	4	2.64
	1	24.4	0.30	.001	2.11	8.6	22	15	7	2.68
	3	25.6	0.30	.001	2.06	9.0	28	13	15	2.58
	5	27.3	0.23	.001	1.98	9.8	37	12	25	2.57
B	2	17.5	0.32	.009	2.11	7.3	26	15	11	2.67
	4	21.3	0.30	.006	2.05	9.0	25	11	14	2.67
	6	22.2	0.28	.004	2.07	8.6	49	18	31	2.53
	8	25.4	0.28	.001	2.02	9.3	63	14	49	2.57
C	0	22.4	0.28	.007	2.12	7.6	15	14	1	2.63
	2	25.3	0.26	.003	2.07	7.8	20	14	6	2.69
	4	26.4	0.26	.001	2.00	8.8	37	14	23	2.65
	6	27.2	0.26	.001	1.98	10.7	27	19	38	2.68

γ_dMax = Maximum Dry Density - T/m^3
ωopt = Optimum Moisture Content - %
LL = Liquid Limit - (%)
PL = Plastic Limit - (%)
PI = Plasticity Index
G_s = Specific Gravity gm/cc

*ASTM D-698 Method C

Twelve test specimens, one each representing a bentonite amended soil mixture having between zero and eight percent bentonite, were fabricated to 95 percent of their respective maximum dry unit weights at a moisture content two percent wet of optimum. Permeability tests were performed as previously described. Results of these tests generally indicated that the target permeability of 5x10^{-8} cm/sec could be attained with the addition of approximately two percent bentonite. The results of these tests are summarized in Table 3 and presented graphically in Figure 2.

Table 3. Phase I Permeability Test Results

Material	Bentonite Percent	Dry Density T/M^3	Moisture Content %	Permeability cm/sec
A	0	1.98	10.7	$3x10^{-8}$
	1	1.99	11.0	$2x10^{-8}$
	3	1.96	11.1	$3x10^{-8}$
	5	1.90	12.1	$1x10^{-8}$
B	2	2.02	9.6	$6x10^{-8}$
	4	1.93	11.0	$2x10^{-8}$
	6	1.96	11.2	$7x10^{-9}$
	8	1.92	10.9	$6x10^{-9}$
C	0	2.01	9.6	$4x10^{-7}$
	2	2.00	10.4	$5x10^{-8}$
	4	1.92	10.6	$1x10^{-8}$
	6	1.90	12.8	$4x10^{-9}$

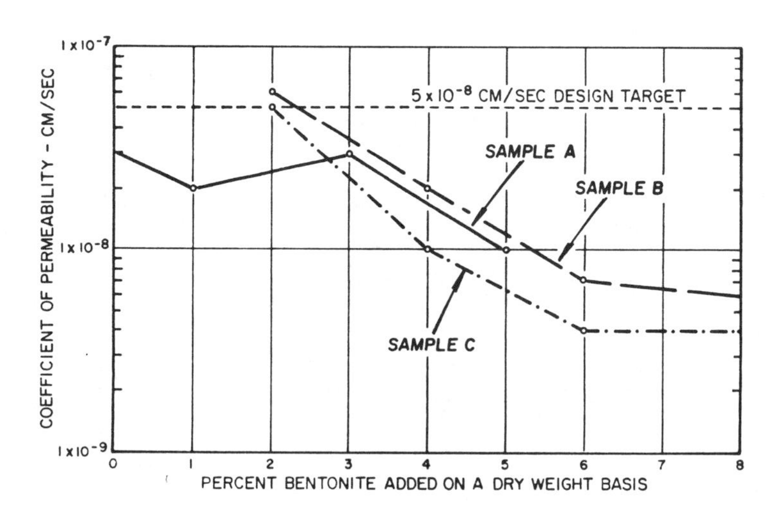

Figure 2. Phase I Permeability Test Results

PHASE II TESTING

Phase II investigated the potential alternation of liner permeability resulting from cyclic freezing and thawing and short-term but continuous unprotected exposure during construction.

In addition, the effects of increased compaction effort on permeability were investigated.

Based on the results of Phase I testing, three subsamples of a composite developed from 10 bag samples were prepared and amended with two percent bentonite. A suite of index and compaction tests were performed on each, the results of which are presented in Table 4.

Table 4 - Phase II Index Test Summary

Sample Designation	Atterberg Limits			G_s	Compaction Data	
	LL	PL	PI		γ_d Max.	ω opt
Bulk Composite (no bentonite)	10	11	N/P	2.7	--	--
S-1 (2% bentonite)	19	11	8	2.7	2.15*	6.8
S-2 (2% bentonite)	19	12	7	2.7	2.14*	7.1
S-3 (2% bentonite)	18	12	6	2.7	2.19**	6.1

* ASTM D-698, Method C
** ASTM D-1557, Method C
γ_d Max = Maximum Dry Density - T/M^3
ω Opt = Optimum Moisture Content - %
LL = Liquid Limit - %
PL = Plastic Limit - %
PI = Plasticity Index
G_s = Specific Gravity - gm/cc

Permeability test specimens were prepared in a similar manner as described for Phase I (i.e., 95% maximum dry density and +2% optimum moisture).

Constant head permeability tests were performed subsequent to freeze/thaw and desiccation exposure on each specimen. The results of these permeability tests are summarized in Tables 5 and 6 and presented in Figures 3 and 4 for the cyclic freeze/thaw and air exposed specimens, respectively.

Table 5. Summary of Results - Phase II Cyclic Freeze/ Thaw Specimens

Specimen No.	S2-A	S2-B	S2-C	S2-D
No. of Cycles	0	2	6	10
Dry Density - T/M^3	2.05	2.03	2.03	2.04
Percent Compaction	94.8	94.7	95.0	95.1
Molding Moisture Content (%)	9.0	9.2	9.3	9.0
% Above Opt. Moisture	1.9	2.1	2.2	1.9
Saturation B Value	1.00	0.96	0.97	0.97
Permeability (cm/sec)	$1x10^{-7}$	$1x10^{-6}$	$4x10^{-7}$	$1x10^{-7}$

Table 6. Summary of Results - Phase II Desiccation Exposure Specimens

Specimen No.	S1-A	S1-B	S1-C	S1-D
Desiccation Time (hours)	0	8	48	96
Dry Density-T/M^3 (before test)	2.05	2.06	2.04	2.03
Percent Compaction	95.1	95.6	95.0	94.6
Molding Moisture Content (%)	8.9	8.8	8.8	9.0
% Above Optimum Moisture	2.1	2.2	2.0	2.2
Saturation B Value	0.99	0.99	0.96	0.98
Permeability (cm/sec)	$6x10^{-7}$	$5x10^{-7}$	$1x10^{-6}$	$2x10^{-7}$

The permeability determined for the specimens prepared based on an ASTM D-1557 compactive effort was $2x10^{-7}$ cm/sec, not significantly different than the permeabilities for the freeze/thaw or desiccation test control specimens.

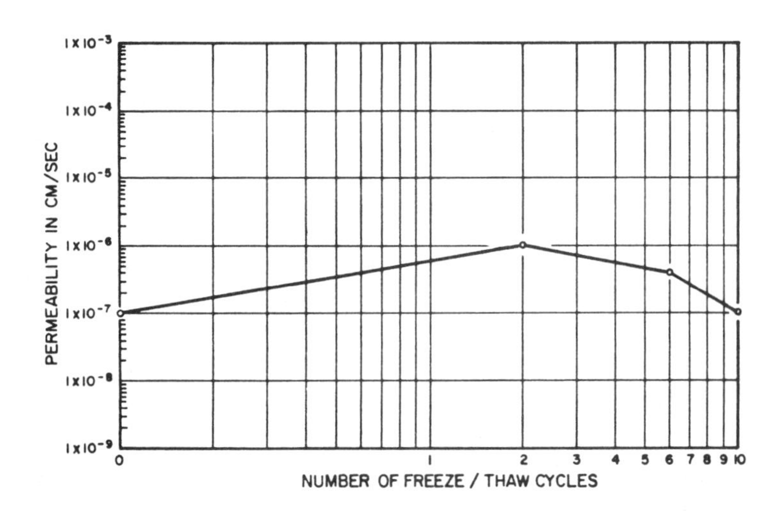

Figure 3. Permeability vs. Cycles of Freeze/Thaw

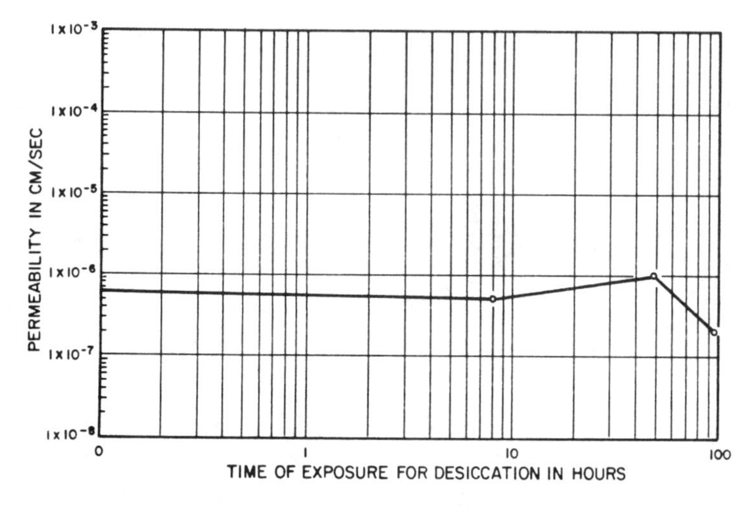

Figure 4. Permeability vs. Hours of Exposure

PHASE III

The results of tests performed on specimens fabricated from the two percent bentonite soil mixture in Phase II indicated difficulty in achieving the target permeability of 5×10^{-8} cm/sec. The Phase III program was designed to reassess combinations of bentonite content and molding moisture required to meet this target.

A 500 Kg composite soil sample was prepared by combining 22 randomly selected bulk samples. This composite was thoroughly mixed by hand and screened to remove the oversize (+19mm) particles. The screened sample was then quartered into subsamples each weighing about 60 Kg.

Subsamples were bentonite amended at their natural moisture content through the addition of 1, 2, 3, or 4 percent bentonite. The four amended subsamples were subjected to index and compaction tests, the results of which are presented in Table 7. Results of grain-size analyses are shown in Figure 5.

Table 7. Phase III Index and Compaction Test Results

Material Designation	% Bentonite	Compaction* γ_d Max.	ω_{opt}	Atterberg Limits LL	PL	PI
III A	1	2.19	6.2	16	15	1
III B	2	2.18	7.0	18	14	4
III C	3	2.15	7.7	21	15	6
III D	4	2.13	8.5	21	13	8

*ASTM D-698 Method C
γ_d Max - T/M^3

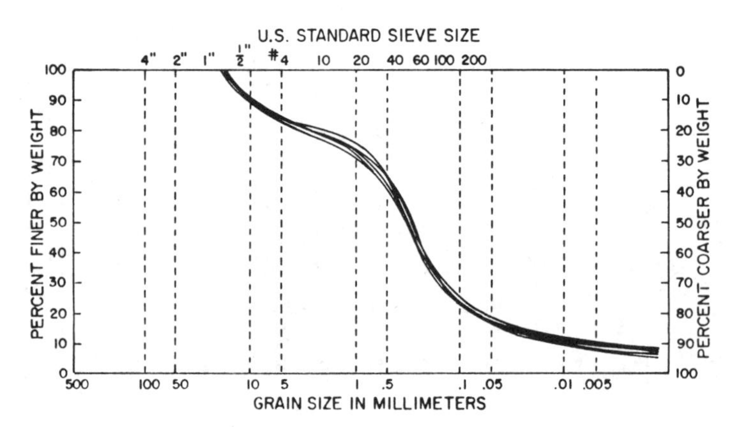

Figure 5. Phase III Grain-Size Curves

Four permeability test specimens were then prepared from each subsample. All specimens were remolded to 95 percent of their respective maximum dry densities at moisture contents ranging from one percent to four percent above optimum. All specimens were consolidated and tested using an effective confining stress of 70 kPa. The results of constant head permeability tests performed on these specimens are presented in Table 8.

Table 8. Phase III Permeability Test Results

Percent	Molding Moisture Above Optimum (cm/sec)			
Bentonite	1 percent	2 percent	3 percent	4 percent
1	$3x10^{-6}$	$1x10^{-6}$	$1x10^{-7}$	---
2	---	$1x10^{-6}$	$4x10^{-7}$	$1x10^{-8}$
3	---	$6x10^{-7}$	$4x10^{-8}$	$1x10^{-8}$
4	---	$6x10^{-9}$	$5x10^{-9}$	$2x10^{-9}$

Long-Term Tests

Two specimens fabricated using the two percent bentonite amended mixture prepared for Phase II were subjected to long-term hydraulic conductivity determinations. Specimen preparation and permeability testing procedures were identical to those used throughout the remainder of this test program. Permeability measurements were taken on a weekly basis. Specimen 2M was tested using a hydraulic gradient of 9 and an effective confining pressure of 35 kPa. Specimen 2S was tested under similar conditions initially. After six months, 2S was consolidated under 240 kPa and tested using a gradient of 55. Test specimen properties are summarized in Table 9. The results of these long-term tests are presented on Figure 6.

Table 9. Summary of Long-Term Test Specimen Properties

Sample	% Bentonite	Dry Density T/m^3	Molding Moisture %	% Compaction
2S	2	2.07	9.0	96.4
2M	2	2.09	8.0	95.4

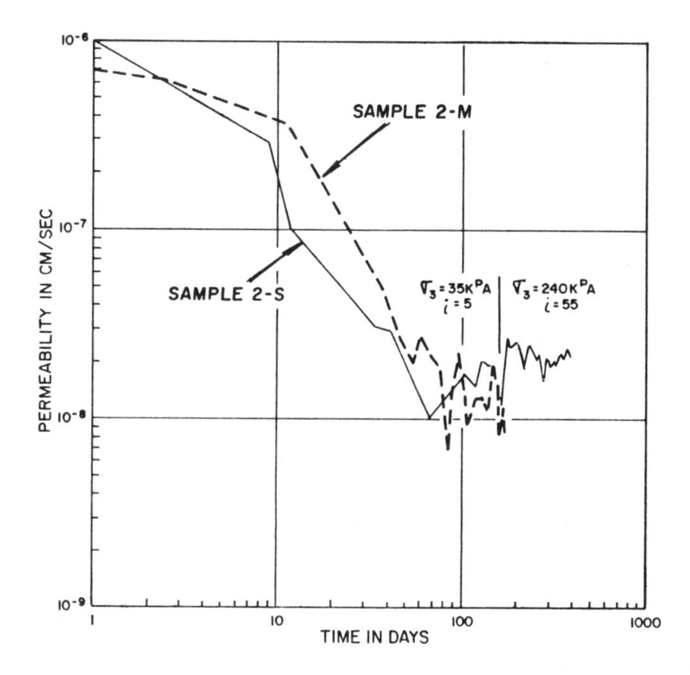

Figure 6. Long-Term Permeability Test Results

CONCLUSIONS

A uniform liner material can be developed from the available on-site materials tested achieving the target design permeability of 5×10^{-8} cm/sec through bentonite amending. The addition of as little as two percent bentonite (dry weight basis) can produce a satisfactory material. Permeability was extremely sensitive to molding moisture content especially at the lower range of bentonite percentages evaluated.

Neither cyclic freeze/thaw or prolonged air exposure up to 96 hours had significant effect on changing permeability values relative to those observed for the control specimens. Variations in permeability both with cycles of freeze/thaw and exposure time were within an order of magnitude and showed no significant negative trends.

The effects of an increased compactive effort on observed permeability had a minimal effect relative to molding moisture content or bentonite content.

The two long-term tests performed in conjunction with Phase II showed a 1-1/2 to 2 order of magnitude decrease in permeability after 60 days of testing with relatively constant values observed thereafter. Initial decreases, those within the first 10 days, are suspected to be associated with delayed bentonite hydration, however, parallel studies exploring pore fluid geochemisty indicate the occurrance of biological sulfate reduction. Pore space plugging associated with this phenomenon may also be contributing to the significant permeability decreases observed between 10 and 60 days.

REFERENCES

1. Knight, R.B., and J.P. Haile, "Construction of the Key Lake Tailings Facility," paper presented at the International Conference on Case Histories in Geotechnical Engineering, St. Louis, MO 1984.

2. D'Appolonia, D.J., "Soil Bentonite Slurry Trench Cutoffs," Journal of the Geotechnical Engineering Division, ASCE, Vol. 106, No. GT4, April 1980

3. Haxo, H.E.,Jr., R.S. Haxo, N.A. Nelson, P.D. Haxo, R.M. White, and S. Dakesaian, Final Report: Liner Materials Exposed to Hazardous and Toxic Wastes. EPA-600/2-84-169. U.S. EPA, Cincinnati, OH 1984.

4. Gipson, A.H.,Jr., "Permeability Testing on Clayey Soil and Silty Sand-Bentonite Mixture Using Acid Liquor," Hydraulic Barriers in Soil and Rock, ASTM STP 874, A.I. Johnson, R.K. Frobel, N.J. Cavalli, and C.B. Pettersson, Eds., American Society for Testing and Materials, Philadelphia 1985.

5. Daniel, D.E., Trautwein, S.J., Boynton, S.S. and Foreman, D.E., "Permeability Testing With Flexible-Wall Permeameters", Geotechnical Testing Journal, GTJODS, Volume 7, No. 3, September 1984.

6. Dunn, R.J. and Mitchell, J.K., "Fluid Conductivity Testing of Fine-Grained Soils", Journal of the Geotechnical Engineering Division, ASCE, Volume 110, No. 11, November 1984.

MULTIPLE-LEVEL GROUNDWATER MONITORING AT THE FMC TRONA PLANT, GREEN RIVER, WYOMING

Wesley K. Nash, Jr
FMC Wyoming Corporation, Alkali Division

Jennifer L. Askey
FMC Corporation, Minerals Division

William Black and Eric Rehtlane
Westbay Instruments Ltd. Canada

ABSTRACT

A multiple-level ground water monitoring system was installed in 1983 at the FMC trona plant near Green River, Wyoming. This system was used to obtain data on the character and movement of ground water in the vicinity of surface processing facilities and waste water lagoons. The monitoring system was installed in 25 drill holes ranging from 90 to 220 feet in depth. Fluid pressures and samples are being monitored at over 140 ports in the 25 drill holes. This paper discusses instrumentation selection, drilling, logging, installation procedures and the piezometric data obtained immediately following installation. Summary pressure data are presented in a representative cross-section showing the geologic units and fluid equipotential lines. The paper concludes with a general discussion regarding the installation and operation of the monitoring equipment.

INTRODUCTION

Project Purpose/Site Description

In conjunction with modifications to an existing tailings impoundment, a ground water study was begun at the FMC trona mine and soda ash processing facility. The facility is located about 17 miles west of the city of Green River, Wyoming as shown in Figure 1.

Geotechnical and Geohydrological Aspects of Waste Management, D. J. A. van Zyl et al., Eds., © 1987 Lewis Publishers, Inc., Chelsea, Michigan – Printed in USA.

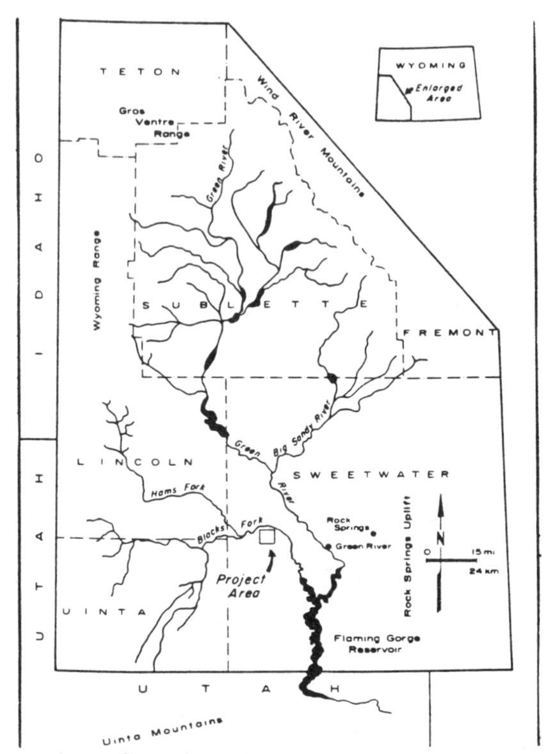

Figure 1. Location of project area

The primary activity is the mining and processing of the mineral trona into soda ash. Processing results in insoluble tailings which are slurried to an impoundment that decants the alkaline slurry water into a large playa lake where evaporation occurs. The waste water impoundments represent a closed system with no surface discharges.

Topographically the site is relatively flat with only 100 to 150 feet of relief. The evaporation lake elevation is about 6280 feet , the tailings impoundment about 6315 feet, the processing plant complex about 6200 to 6250 feet, and the Blacks Fork River and alluvial flood plain about 6180 to 6190 feet. Figure 2 shows the general layout of the facility.

Average annual precipitation at the site is 6.9 inches, and average annual evaporation is 42 inches making the area semiarid to arid. Temperatures range from a high of 93 to a low of -45 degrees F. with an annual mean of 45 degrees F.

Preliminary Work

Early in the project aerial photographs, topographic maps, and visual inspections were used to determine probable sources and (more importantly) routes for seepage. Shallow seismic surveys and exploratory drillings were then conducted across these probable routes to further define their subsurface features.

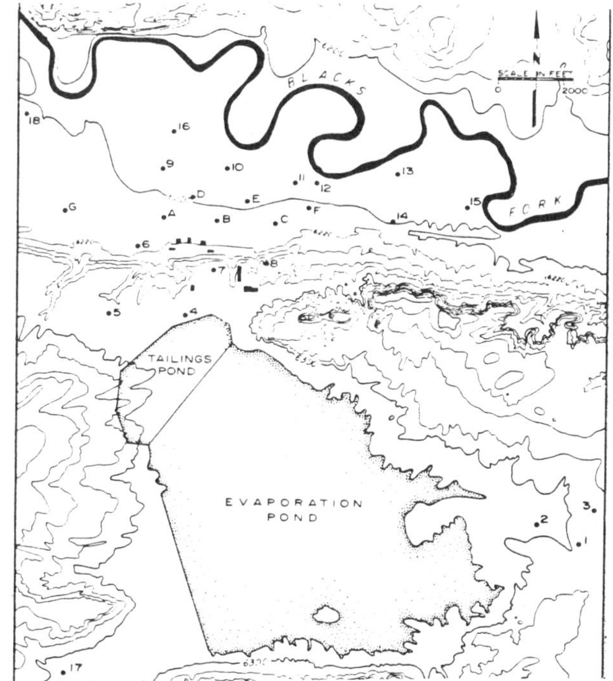

Figure 2. Study site and monitoring well location.

SELECTION OF MONITORING METHOD

Role Of Geology

The site is situated atop sediments of the Eocene Bridger Formation and recent alluvial deposits of the Blacks Fork River. Information from exploration work indicated that several water bearing zones with significantly different hydraulic characteristics existed. These zones were in the Bridger Formation and the recent alluvium of the Blacks Fork River. The Bridger Formation is composed primarily of very fine-grained silty-sandstone and clayey shale with lenses of clay and limestone. The Blacks Fork River alluvium consists of interbedded sand, silty clay, and gravel. Three oil shale layers in the study area served as zonal marker beds.

These site-specific characteristics, which govern the hydrogeological regime, played an important role in the choice of the ground water monitoring system. Five to six monitoring zones were targeted at each location for flow regime and water quality monitoring. This array permitted both the principle geologic units and the conventionally low hydraulic conductivity units to be instrumented. These secondary units could behave either as aquitards or as fracture-flow aquifers and, therefore, monitoring them was deemed important.

A reliable, cost-effective three-dimensional ground water monitoring system was required. The selected system had to allow for the collection of fluid samples, fluid pressure data, and formation hydraulic characteristic testing. At the same time the system had to be easily installed, operated and maintained at the proposed depths plus be capable of periodic verification as part of an overall quality assurance program.

As part of the selection process, a cost and technical comparison was made between a clustered standpipe piezometer monitoring system and the multilevel, single hole, Westbay MP monitoring system. As illustrated in Figure 3, a clustered standpipe system consists of a series of adjacent drill holes each with a standpipe completed at a different depths. The MP system described by Black et all [1], provides multiple level monitoring in a single drill hole.

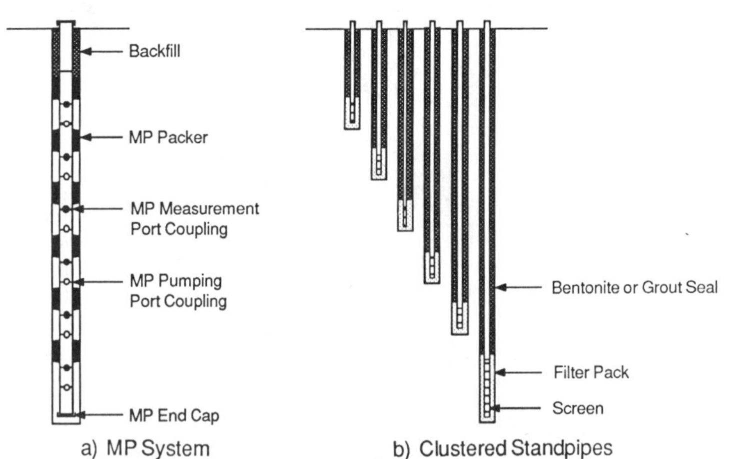

Figure 3. Typical MP and standpipe multiple level installation.

Technical Comparisons

Number of Drill Holes Required

The purpose of installing an individual ground water monitoring zone is to characterize and monitor the hydrogeologic and hydrogeochemical conditions at a given point over a given time. Although drilling provides direct access to a particular point, it may alter the natural state of the ground water regime. Also, the introduction of backfill or other foreign sealing materials adjacent to a monitoring zone may alter the sample chemistry. Clustered standpipes require individual drill holes, each of which involves a hydraulic disturbance that increases the potential for adverse effects. The MP system requires a single drill hole for multiple monitoring zones, thus substantially reducing the potential for disturbing the natural hydrogeologic system.

Drill Hole Annulus Seal

Clustered standpipes require a graded filter material to be placed so as to fill the annular void surrounding the intake screen. The remainder of the annular space surrounding the standpipe should be filled with a sealing material. This procedure is difficult, especially in deeper holes such as those associated with this site. Considerable care must be taken to place the seals in the correct position to prevent backfill smearing in monitoring zones and to prevent the migration of sealing materials into adjacent standpipe drill holes. Large vertical hydraulic gradients make these operations particularly difficult.

The MP system is installed using casing packers filled with water to seal the annulus between monitoring zones. Numerous packers can be located in a single hole. Each packer is individually inflated with the inflation pressure and volume carefully controlled. Any uncertainty can be checked and, if necessary, the packer can be further inflated. Packers are inflated immediately after drilling thereby reducing contamination from flow within the open drill hole. Back-fill materials may also be used for annulus seals with the MP system.

Fluid Pressure Measurements

In low hydraulic conductivity formations, the MP system responds more quickly to pressure variations than open standpipes because almost no fluid displacement takes place, whereas significant displacement of fluid occurs in open standpipes.

Pressure response testing can be part of every pressure measurement in a standpipe monitoring well thereby verifying that the inlet screen and riser are behaving consistently. The same verification is possible with the MP system.

Typical water level measurements in standpipe piezometers are made with tapes or electrical transducers lowered into the standpipe to locate the top of the water column. With the MP system, fluid pressures are measured by means of a transducer inside a probe. The pressure probe is lowered into the MP casing and located adjacent to the monitoring zone. The transducer is connected to the outside formation through a valve in the wall of an MP coupling. Pressure and temperature data appear on a readout unit located at the surface. Because the pressure probe with it's transducer is removed after each set of measurements, maintenance and calibration checks are easily accomplished. An added feature of the MP system is that fluid pressures below atmospheric can be measured.

Ground Water Sampling

Access to formation water in a standpipe system is through a well screen. The water enters the standpipe and stands in a column above the screen. The water is in a different physical, biological and chemical environment from that in the formation and samples are influenced by this water column. For this reason, it is standard practice to purge the riser pipe by flushing formation water through it. Typically, three to five well volumes might be removed. This purging has several repercussions. Besides requiring excessive amounts of time, purging produces pumped water that is sometimes considered a pollutant requiring costly handling and disposal procedures. Pumping lowers the fluid pressure around the well screen, imposing unnatural fluid pressures and fluid flow paths. These conditions may introduce water to the riser which is not representative of the formation water at the monitoring zone. The purging may increase the turbidity of the water entering the well bore thus degrading the sample quality.

The MP system minimizes these problems since samples are collected from outside the MP casing. Fluid inside the casing is isolated from the formation during sampling and will not influence sample quality. Purging, although possible at any time, is not required except immediately after installation when it is used to remove the fluids that were affected by drilling.

Equipment used to collect samples in standpipes should be decontaminated between samples. Procedures similar to those used for purging are appropriate if dedicated sampling equipment is used. Tubing or cable surfaces contacting the sampled water should be thoroughly cleaned between samplings, and when long lengths of tubing or cable are involved such decontamination can be time-consuming and expensive. Only with care and much difficulty will cross-contamination of samples be avoided, if at all.

With the MP system, only the inside of the sampling probe and the inside of the sampling container require decontamination. This greatly simplifies procedures, decreases field time, and reduces the consumption of cleaners and/or distilled water. The small surface area exposed also reduces the probability that serious cross-contamination will occur should decontamination be inadequate.

Sample collection with the MP system is easy and can be accomplished using a variety of techniques. Sampling procedures can be varied depending on site conditions and expected ground water chemistry. Duplicate samples are easily collected.

Hydrogeologic Testing

A standpipe system permits many hydrogeologic tests to be carried out including variable head tests, tracer injection tests, and point dilution tests. The MP system also allows all these tests to be performed. The modular nature of the MP system allows additional tests to be performed, several involving quality assurance, which are not possible in a standpipe system.

Quality Control

Field quality control involves well installation, monitoring equipment and sampling procedures. Some of the quality control checks built into the MP system are difficult or impossible to duplicate with clustered standpipe systems. For example, the hydraulic integrity of a monitoring well's riser pipe is critical. The interior of the MP casing is hydraulically isolated. Water levels can be expected to remain stationary and can be adjusted to be different from formation water levels. As a result, the casing integrity can easily be checked at any time. With a standpipe system, internal packers must be placed above the screen, and the riser pressurized each time hydraulic integrity confirmation is required.

Confirmation of annulus sealing is another monitoring well quality assurance requirement. Direct testing across the seal, the most positive test, is not possible in a standpipe system. With the MP system, pressure may be fluctuated on one side of a seal and the response monitored on the other. Packers can also be checked and the packer reinflated if necessary.

Pressure measurement quality control is not easy with a standpipe system. Accurate tape length checks require tensioning to duplicate field measurement conditions. In a standpipe the water column inside the casing and above the screen is part of the measurement system. Therefore, periodic density checks may be needed, especially in deep wells.

Water of known density can be maintained within the MP system casing. At each monitoring zone, pressure applied by inside casing water can be used for field calibration checks. Transducer output can also be compared against laboratory standards. Although the MP pressure probe is more complex than a water level tape, normal operating procedures offer field indications of whether the probe is operating correctly.

Sample blanks and spikes, vital to a good quality assurance program, are possible in a standpipe system when a down hole pump is used but very difficult with bailers. With dedicated pumps, spikes can be impossible to obtain. The MP system's sampling probe easily reproduces the down hole sampling procedure with all critical surfaces exposed to the blanks and spikes.

Cost Comparisons

A cost comparison was prepared in 1983 using prices from Westbay Instruments Ltd and from two separate competitive PVC monitoring well casing manufacturers. The unit price of the MP system casing components was found to be higher than that of plain PVC well casing. However, as noted by Rethlane and Patton [2], if data is required from more than two or three different depths at each site, the MP system becomes cost effective due to the reduction in drilling. This also has significant effect on project schedule since drilling is usually a major item.

While the cost determinations are too lengthy for the scope of this paper, the comparisons carried out suggested the use of the MP system would result in costs on the order of 60% lower than the cost of clustered standpipes. Table 1 gives a comparison of the approximate total cost using actual figures for the MP system and estimated figures for standpipes.

Selection of Monitoring System

The significant technical and economic advantages of the MP system made it's choice apparent. However, this system had never been used in Wyoming. Since the data obtained would be reviewed by the State, their approval was needed. Prior to any purchase commitment by FMC, Westbay personnel participated in a meeting held with the State which lead to the necessary approval.

Table 1. Total Cost Comparison MP vs. SP System

Item	MP System	SP System
Number of Holes	25	150
Number of Monitoring Zones	150	150
Time for Installation	4 weeks	12 weeks
Total Cost	$300,000	$900,000

INSTALLATION

Planning and Preparation

Once State approval was received, a predrilling meeting was held with the drilling contractor (Lang Exploratory Drilling), field personnel, and the FMC geologist and environmental engineer. The purpose of the meeting was to discuss the proposed drilling method and to identify any potential problems that might occur during installation. Contingency plans were developed for each problem identified. This planning contributed to the overall success of the construction phase of the project.

Drilling Method

Lang Exploratory Drilling was contracted to drill the monitoring wells using a dual-tube reverse circulation air rotary rig. Drill cuttings were fed to a rotating splitter to prevent sample cross-contamination. This method provided an opportunity for the geologist to construct graphic logs of each hole as drilling progressed. The method also minimized the disturbance of formation materials away from the borehole wall since the cuttings are transported to the surface through the drill stem and not up the annulus. The technique was very successful, resulting in a highly uniform hole with side-wall clearances well within the tolerance limits of the inflatable packers.

Geologic Logging

Cuttings were collected at five foot intervals, washed, and glued to a chipboard for a permanent record. Monitoring horizons were selected on the basis of the chip samples and geologist's description. Monitoring ports were placed in sandstone, siltstone, and limestone horizons as well as the weathered Bridger horizon and the Blacks Fork River alluvium. Rough cross-sections were drawn in the field and used to check the congruency of individual well monitoring horizons. The oil shale layers previously mentioned provided an excellent means of cross-checking the stratigraphic units.

A concern raised in the predrilling meeting was whether casing placement and packer inflation work would proceed so rapidly that the installation personnel would be waiting on the driller or geologist. This was not the case. Prior exploration and careful logistical planning resulted in a well coordinated field operation.

Field Installation of the MP System

After the driller completed a hole and had moved to another site to begin the next hole, the FMC environmental engineer, in consultation with the geologist, designed each well. This involved chosing packer and port locations and the number of ports needed. The necessary MP casing components were then selected, labeled according to their proper placement, and laid out at the drill hole. MP casing components were successively attached to one another and lowered into place. Since the MP casing is a sealed system, it was buoyant in drill hole water. Clean water was added to inside the MP casing to neutralize the buoyancy.

At each monitoring zone a 20 to 80 micron polyester geotextile filter was placed around measurement and pumping port couplings. These were to reduce the possibility of fines clogging or interfering with the action of the valved couplings.

Packers were inflated immediately, with individual inflation volume and pressure being recorded to verify performance.

Pressure measurements were taken as soon as practical to verify the operation and performance of installed equipment.

Surface completion of each well consisted of a steel surface casing sealed in the drill hole annulus. A lockable steel box welded to the top of the casing provided access and served as a convenient location for a log of the instrumentation in that well.

RESULTS

Cost and Schedule

The ground water monitoring network installation cost came in under the estimated budget and was completed ahead of schedule. On average one 150 foot well system was completed per day. Much of the success was due to prior planing, efficient drilling and installation crews and a monitoring system that was easily adapted to the different project requirements.

Initial Pressure Readings

Initial fluid pressure readings indicated that all the instrumentation strings were successfully installed and operating properly. The initial pressure readings, along with geologic interpretation enabled preliminary interpolation of fluid equipotential lines on cross-sections such as that shown as Figure 4. The equipotential indicate a downward flow from the recharge area in the upland near the tailings impoundment and upward flow to discharge in the river. The data also suggests that a three-dimensional monitoring system was the proper choice for this site.

Operating History

The system has been operational for three years. Over 1500 samples have been collected, over 2000 pressure measurements have been taken, and over 40 variable-head tests for hydraulic conductivity have been performed. During this period no scheduled sample runs have been missed due to equipment failure. The only problems encountered were easily resolved through minor equipment modifications. The data are being used effectively for predictive modeling projections that will be used in a report to the state regulatory agency.

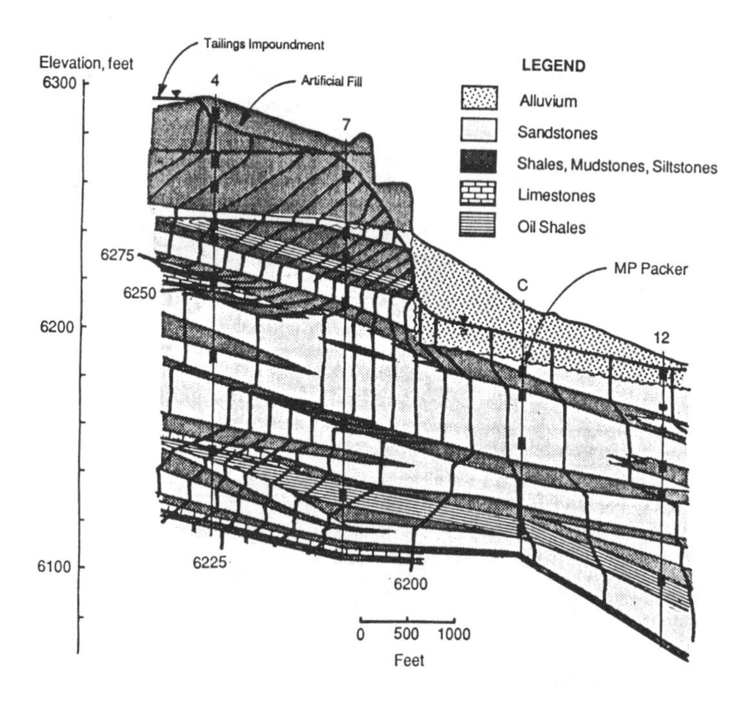

Figure 4. Hydrogeologic section, showing major geologic units, and initial fluid equipotential lines.

CONCLUSION

The particular hydrogeological environment encountered at this site required a three-dimensional ground water monitoring network. The system also needed to be compatible with the particular drilling technique chosen as well as simple, quick and efficient to install and operate. The MP system was uniquely suited to these needs.

The predrilling conference developed contingency plans that were used effectively for several problems that arose. This early planning also ensured that proper equipment was on hand, thus preventing delays.

The drilling technique chosen provided a clean drilling method that minimized borehole wall damage and intrusion of fluids into the formation as well as providing good chip samples for use in geologic logging.

The MP system was installed simply and quickly, and continues to provide accurate fluid pressure data and high quality fluid samples, as well as allowing hydraulic conductivity testing. It was the most economic and schedule-effective alternative for ground water monitoring at this site.

1. Black W.H., Smith H.R., and Patton F.D. "Multiple-Level Ground Water Monitoring With the MP System," Surface and Bore Hole Geophysical Methods and Ground Water Instrumentation, A Conference and Exposition Denver, Colorado, October (1986).

2. Rethlane E.A., and Patton F.D. "The MP System vs. Standpipe Piezometers: An Economic Comparison", Second International Symposium on Aquifer Restoration and Ground Water Monitoring, Columbus, Ohio, May (1982).

DESIGN OF GROUNDWATER MONITORING SYSTEMS—HYDROGEOLOGIC CONSIDERATIONS

Edward Mehnert, Illinois State Geological Survey
and the Hazardous Waste Research and Information Center, Champaign, Illinois

Beverly L. Herzog, Bruce R. Hensel, Jerry R. Miller,
Illinois State Geological Survey, Champaign, Illinois

Thomas M. Johnson, Levine-Fricke Consulting Engineers
and Hydrogeologists, Oakland, California

INTRODUCTION

The design of groundwater monitoring systems depends on many factors. A major factor is the scope of the monitoring system, i.e., what is the relative size of the area to be monitored. Everett [1] presents guidelines for developing and implementing regional groundwater quality monitoring systems. Nacht [2] discusses several generic factors important to the proper design and operation of groundwater monitoring systems. Nacht describes a number of factors common to most groundwater monitoring systems, including objective(s) of monitoring, number of samplers, and selection of an analytical lab, which, if not adequately considered, could lead to improper interpretation of data.

This paper focuses on the design of groundwater monitoring systems for hazardous waste management facilities (landfills, surface impoundments, land treatment facilities, etc.). Groundwater monitoring systems for these facilities must comply with the regulations put forth in the Resource Conservation and Recovery Act (RCRA) of 1976 and the Hazardous and Solid Waste Amendments of 1984. Specifically, results of a study, conducted by the Illinois State Geological Survey (ISGS), evaluating the groundwater monitoring systems for several, existing hazardous waste management facilities in Illinois are presented. Hydrogeologic considerations for the design of groundwater monitoring systems, in general, are also discussed.

Geotechnical and Geohydrological Aspects of Waste Management, D. J. A. van Zyl et al., Eds., © 1987 Lewis Publishers, Inc., Chelsea, Michigan—Printed in USA.

GROUNDWATER MONITORING REGULATIONS

Groundwater monitoring for RCRA hazardous waste management facilities is subject to numerous federal and state regulations. The following discussion emphasizes those regulations pertaining to the design of groundwater monitoring systems. Regulations concerning the design of monitoring wells will not be discussed. Since regulations are written to apply to sites across the country, the regulations generally specify the minimum requirements for the design of groundwater monitoring systems. For a particular site, a more complex groundwater monitoring system is usually necessary to adequately monitor groundwater flow and quality.

Graves [3] discusses the groundwater monitoring requirements of RCRA prior to the enactment of the 1984 amendments to RCRA. Graves, who was involved with drafting the groundwater regulations of RCRA, states that protection of groundwater quality was a major objective of the RCRA program. Under RCRA, groundwater monitoring systems fall into three categories: 1) for a facility known or assumed not to be contaminating groundwater (detection monitoring); 2) for a facility known or assumed to be contaminating groundwater (assessment monitoring); and 3) for a facility which can justify an alternative groundwater monitoring system due to a low potential for migration of hazardous waste constituents from the facility. Sites in the first category would be required to establish a detection monitoring program. Sites in the second category would be required to establish an assessment monitoring program, which is generally a more stringent monitoring program (i.e., these sites may be required to install more monitoring wells and/or analyze for a greater number of chemical parameters) than detection monitoring. Sites in the third category would be required to establish a less stringent monitoring program than a detection monitoring program. In fact, certain sites may not be required to have any groundwater monitoring wells. For each of the categories, Graves outlines in detail, the monitoring and reporting requirements of RCRA, prior to the enactment of the 1984 amendments.

Additionally, federal regulations (see 40 CFR Part 264, Subpart F) require that a "sufficient number of wells" be installed to monitor the "uppermost aquifer." Monitoring of upgradient and downgradient groundwater quality is also required by federal regulations. Several years ago, "a sufficient number of wells" was interpreted to mean one upgradient and three downgradient monitoring wells. Presently, Illinois groundwater monitoring regulations (see Title 35, Subtitle G, Chapter I, Subchapter c, Part 725, Subpart F) specifically state a minimum of one upgradient and three downgradient monitoring wells. The 1984 amendments to RCRA did not change the language concerning the minimum number of monitoring wells; however, the amendments did eliminate exemptions from groundwater monitoring for certain types of hazardous waste management facilities.

ILLINOIS STATE GEOLOGICAL SURVEY STUDY

The ISGS conducted a legislatively mandated study to evaluate the groundwater quality and groundwater monitoring systems in the immediate vicinity of hazardous waste management facilities [4]. From approximately 170 land-based, hazardous management facilities in Illinois, ten facilities were selected for this study. The selection of these sites was based on the following criteria: history of the operating groundwater monitoring system; type of geologic and geographic setting; and, type of hazardous waste management facility (landfill, surface impoundment, etc.). Sites with longer histories of groundwater monitoring were preferred over sites with shorter histories of groundwater monitoring. Sites were also selected in order to achieve a balance in the type of geologic and geographic setting found in Illinois and the type of hazardous waste management facility. Figure 1 shows the locations of the facilities evaluated in this study.

Information for the site evaluations was obtained from the Illinois Environmental Protection Agency (IEPA) and ISGS records. The majority of site specific data was obtained from permit applications and other information filed with IEPA. For certain sites, ISGS records also contained site specific information. Regional geologic and hydrogeologic data were generally obtained from ISGS publications, maps and records. In addition, site operators provided varying amounts of information and commented on draft versions of the ISGS report. Eight of the ten sites were visited by ISGS personnel. During site visits, site personnel were interviewed to clarify information obtained from state records and to obtain additional information.

Based on the evaluations of the groundwater monitoring programs at the hazardous waste management facilities, Herzog et al. [4] made five basic recommendations:

1) Thorough hydrogeologic studies are necessary before an adequate groundwater monitoring program can be established. The depths, extent, and hydrogeologic properties of the geologic materials comprising the shallow groundwater system, as well as the direction(s) of groundwater flow must be understood to properly determine the placement of monitoring wells. This information should be obtained from a detailed hydrogeologic investigation prior to design of the monitoring system.
2) The design of a groundwater monitoring system should consider each site individually. Sites with more complicated geologic conditions will probably require more monitoring wells. Details such as well spacing and screen length should be designed according to the subsurface conditions at the site.
3) Good record keeping is needed for efficient evaluation of waste disposal practices and the effects of these practices on groundwater.
4) Greater use should be made of geophysics, computer modeling and other techniques for efficient and economical site characterization and groundwater monitoring. In particular, geophysics and computer modeling could be used more effectively to implement and supplement traditional monitoring practices.
5) Since records documenting the monitoring programs are scarce for on-site waste generators and disposers of hazardous waste, it appears that more attention should be focused on these facilities. Too little information exists for these operations, despite the fact that many of these facilities may pose a significant hazard.

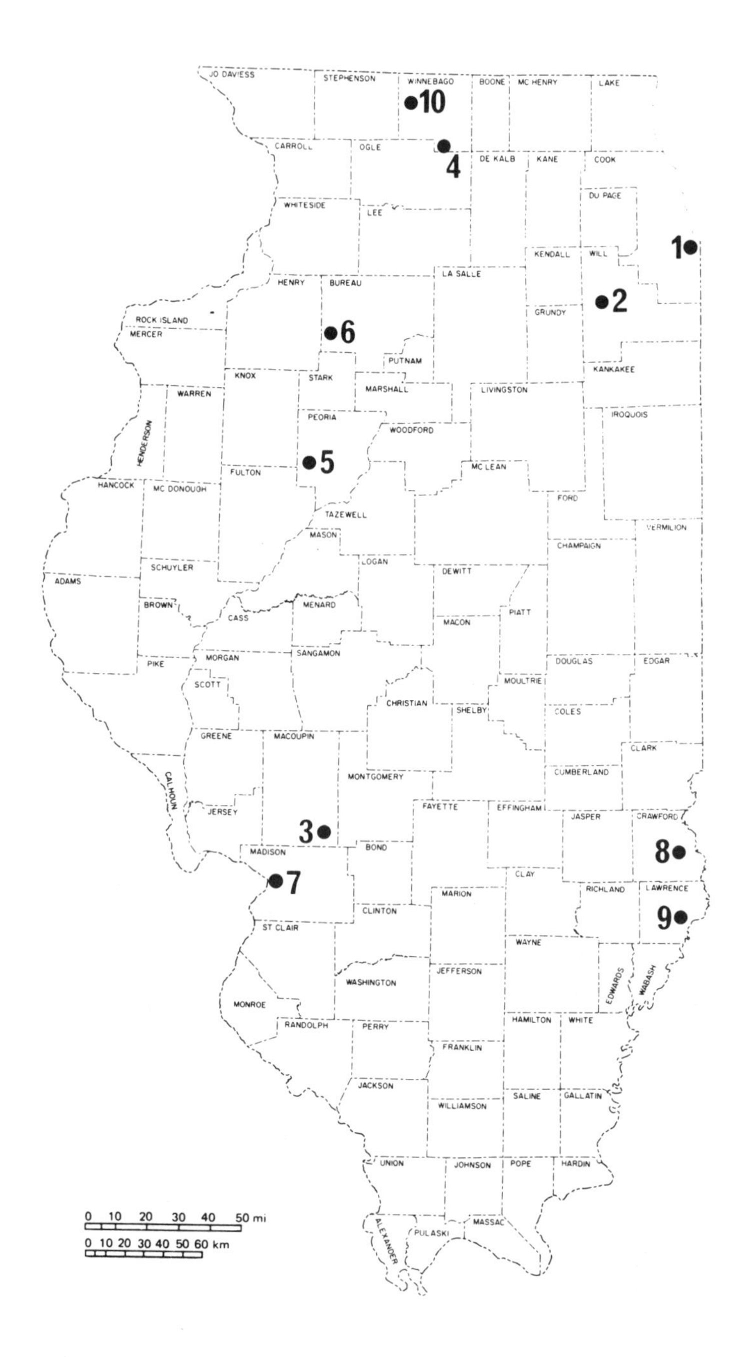

Figure 1. Location map for ten hazardous waste disposal facilities in Illinois selected for evaluation by Herzog et al. [4]. Sites described in this report are Sites 6, 9, and 10.

The remainder of this paper will elaborate on the first and second recommendations. These recommendations were made because nine of the sites evaluated initially had inadequate groundwater monitoring systems. These systems were considered inadequate since their designs were based on inadequate characterization of the site hydrogeologic setting.

Herzog et al. [4] discussed groundwater monitoring systems at ten hazardous waste management facilities. This paper will describe the groundwater monitoring systems at three of these sites. These case histories will illustrate the need to plan and construct groundwater monitoring systems more carefully.

CASE HISTORY 1 (Site 6)

Site 6, a hazardous waste landfill in northwestern Illinois (see Figure 2), received its operating permit from IEPA in 1974; closure is pending. The site is located on top of a hill, and is underlain by a complex sequence of glacial deposits that include approximately 15 meters (50 feet) of loess, sand (a local aquifer) and till. The uppermost bedrock is shale of Pennsylvanian age. Wastes disposed at this site were containerized before landfilling; they include include solvents, neutral organics, organic acids, cyanides, pesticides, reducers, oxidizers, acids and bases.

In order to characterize the geologic and hydrogeologic conditions of the site, six test borings were drilled in 1973. Five monitoring wells were installed before receipt of the operating permit from IEPA in 1974. The site plan is shown on Figure 2, which also shows well locations. The exact locations for three of these wells were specified by IEPA. Due to the quality of the descriptions recorded in the boring/well logs, the geologic material in which these wells are completed could not be accurately determined. In the following years, additional wells were added. In 1976, eleven wells were installed 60 to greater than 180 meters (200 to greater than 600 feet) feet south of the waste site, in conjunction with an adjoining, but separate, waste disposal facility. Two more monitoring wells were installed west of the site in 1977 and three north of the site in 1978. In 1980, sixteen more wells were added increasing the number of on-site wells to 37.

In 1982, contamination was detected in two wells immediately south of one of the disposal trenches. In response, thirteen wells were installed. Data from these wells indicated that organic contaminants were present in an area covering approximately 0.6 hectares (1.5 acres). Most of the contamination was found in wells immediately south of Trench 18EWC. These wells were completed in a sand layer, which had been interpreted as a thin and discontinuous sand layer during the original site hydrogeologic description in 1973.

Eleven wells have been added at the site since 1982 for a total of 72 monitoring wells. The location of the majority of these wells is shown on Figure 2, which also indicates the type of geologic material where the wells are finished. Since the site is situated on a groundwater high, there are no true upgradient wells.

This case history demonstrates the importance of the initial hydrogeologic site description. One can only speculate whether better descriptions of the geologic materials, as recorded on the original boring and well logs, could have led to a more accurate interpretation of the site hydro-

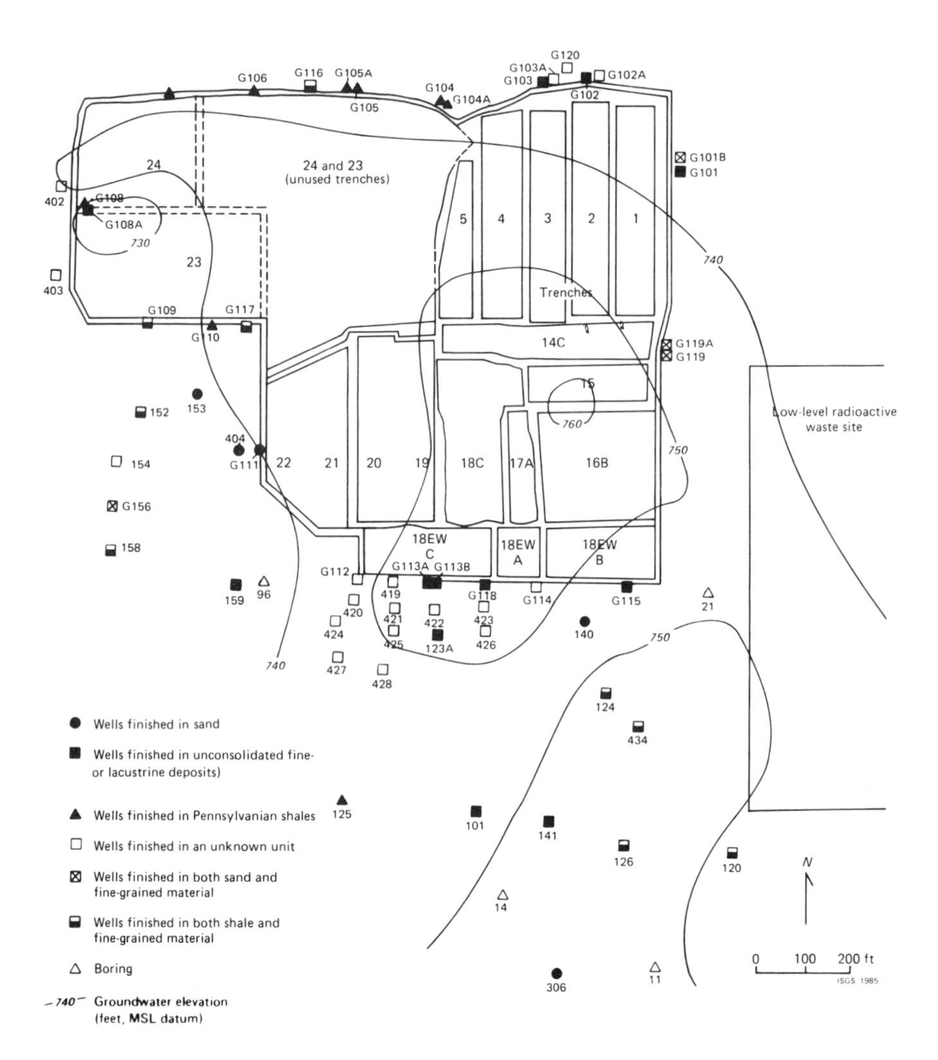

Figure 2. Map of Site 6 showing disposal areas, well locations, and contours of the top of the saturation zone (contour data from 1984).

geology. If the sand zone had been properly characterized during the initial site description, this zone probably could have been more thoroughly monitored. Proper monitoring would have detected contamination at an earlier time, thus reducing the areal extent of contamination prior to detection. In summary, this example illustrates that the initial hydrogeologic description for a site must be complete and accurate in order to properly design the groundwater monitoring system for that site.

CASE HISTORY 2 (Site 9)

Site 9 in southeastern Illinois, is an oil refinery which operated a land treatment facility and surface impoundments before closing in 1985 (see Figure 3). The land treatment facility was used to treat API separator sludge and other sludges derived from refinery operations. Corrosive liquid wastes were managed at the surface impoundment. These hazardous waste management facilities were operated under a RCRA Part A permit. An application for a Part B permit had been submitted. The refinery is situated adjacent to a river. The site consists of an upland area and a lowland area divided by a 3-meter (10-foot) bluff, which trends northeast-southwest. The land treatment facility is located in the upland, while the surface impoundments are located in the lowland.

A geologic cross section of the lowland area is shown on Figure 4. The lowland is characterized by recent alluvial deposits overlying glacial outwash sands and till. The till, not shown on Figure 4, underlies the outwash sand. The geologic deposits underlying the upland are loess, lacustrine and alluvial deposits (interbedded sands, silts and clays) and till. Sandstone of Pennsylvanian age underlies the till. Depth to bedrock averages 7.6 to 15.2 meters (25 to 50 feet) across the site and increases toward the river. The principal aquifers in the vicinity of the site are sand and gravel lenses in the till, which are used for domestic supplies and the outwash sand and gravel deposits in the river valley, which are used for municipal supplies.

Groundwater monitoring near the hazardous waste management facilities began in 1982 with the installation of ten monitoring wells. In the lowland (near the surface impoundments), the zone monitored has primarily been the glacial outwash sand at a depth of 9.1 to 10.7 meters (30 to 35 feet). In the upland area (near the land treatment facility), the interbedded alluvial sand has been the zone monitored at a depth of 4.5 to 6.0 meters (15 to 20 feet).

In order to correct four deficiencies in the original groundwater monitoring system identified by IEPA, six additional monitoring wells were installed by the site operator in 1984. These wells were installed to allow: 1) determination of more realistic background groundwater conditions; 2) detection of possible contamination in a "perched" water zone beneath the surface impoundments; 3) determination of the vertical gradients near the surface impoundments; and 4) detection of possible contamination in an area west of the land treatment facility which had not been previously monitored.

This case history illustrates the need to design a groundwater monitoring system based on site specific, hydrogeologic conditions. Two years after the installation of the initial monitoring wells, additional monitoring wells were installed in order to correct deficiencies in the original monitoring system. At this site, the original groundwater monitoring

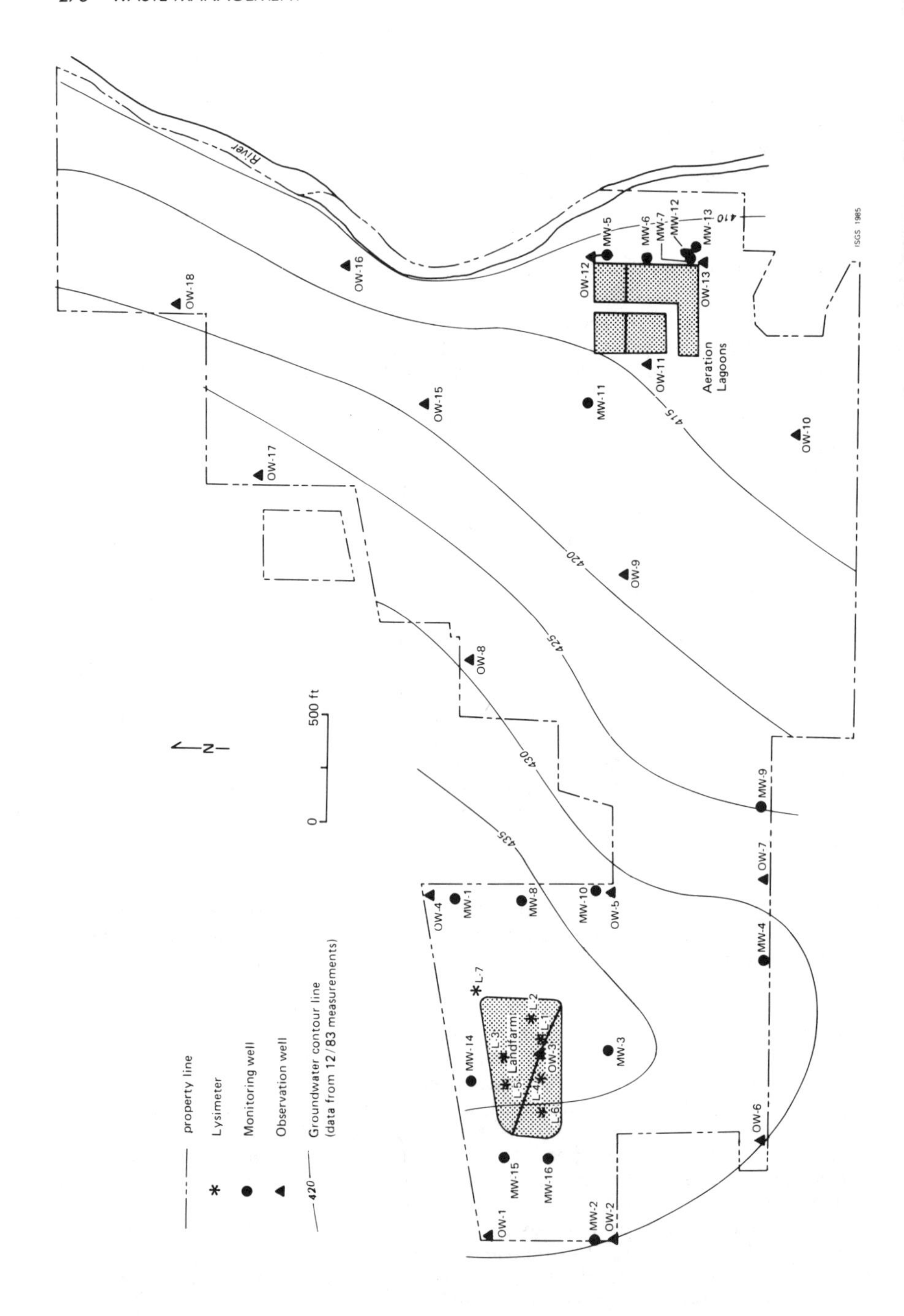

Figure 3. Map of Site 9 showing disposal areas, well locations, and the water table.

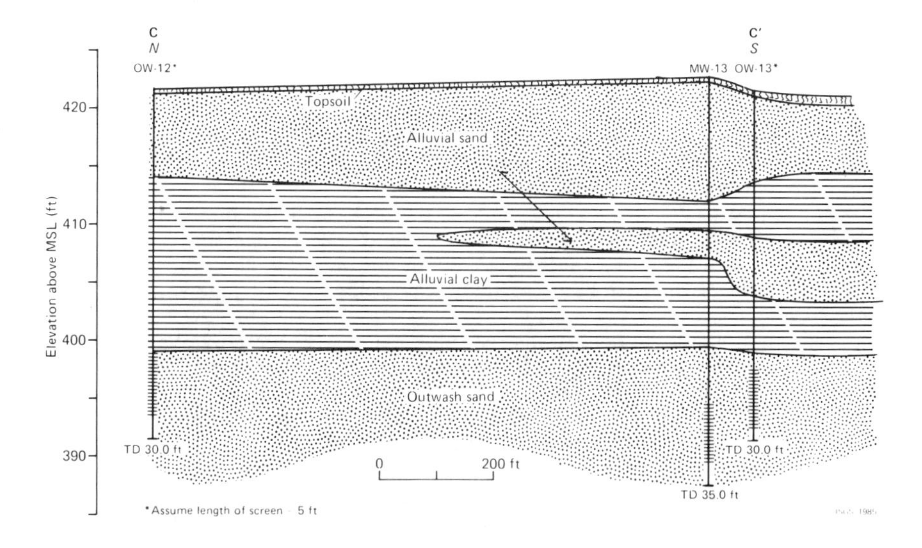

Figure 4. Cross section of Site 9 from north to south, east of the surface impoundment.

system did not monitor the most probable pathways for contaminant transport in the uppermost aquifer. If contamination migrated along these pathways, it probably would have gone undetected for at least two years.

This case history also illustrates the need to monitor all geologic formations capable of transporting contaminants, not just the uppermost aquifer as defined by RCRA regulations. At this site, there are permeable zones above the zone determined to be the uppermost aquifer, which are capable of transporting contaminants. Although groundwater contamination at this site has never been confirmed, the potential exists for significant quantities of leachate to flow through these unmonitored permeable zones, possibly to the river.

CASE HISTORY 3 (Site 10)

Site 10, located in northern Illinois, is a tank storage facility which began operations in 1975 and was permitted in 1979 by IEPA for the storage of liquid hazardous waste (see Figure 5). Liquid wastes had been stored onsite in 17 underground storage tanks and 2 surface impoundments. The storage capacities of the tanks ranged from 11.4 to 79.5 cubic meters (3,000 to 21,000 gallons) and the storage capacity of the surface impoundments was approximately 757 cubic meters (200,000 gallons) each. Wastes stored at this site have included waste oils, metal sludges and solvents.

The hydrogeologic conditions for this site were reportedly characterized from the driller's log of the site's water well. Later geologic investigations have shown that loess and sandy till comprise most of the 6.1 to 18.3 meters (20 to 60 feet) of unconsolidated deposits overlying the bedrock. The uppermost bedrock unit is weathered dolomite of Ordovician age. This dolomite unit is an important local aquifer and is designated as the uppermost aquifer for RCRA monitoring purposes.

The supplemental construction permit issued by IEPA in 1982 required the installation of four monitoring wells—three along the south boundary and one along the north boundary of the site. The wells along the south boundary were apparently believed to be downgradient. The actual groundwater gradient was not known since the site only had one monitoring well prior to the installation of these four wells. The original monitoring well was located approximately 52 meters (170 feet) south of the site. The 1982 wells were drilled to a depth of about 18 meters (60 feet). Three of these wells were completed in the dolomite; the fourth well was completed in the till. This fourth well was later redrilled and finished in the dolomite. Contamination (organics) was reportedly detected during or shortly after the installation of these four wells. Contamination was found in one of the four wells completed in the dolomite aquifer and subsequently in domestic wells located 0.8 kilometers (1/2 mile) south of the site. These domestic wells were also completed in the dolomite.

Additional wells were installed to further define the contaminant plume. In 1983, three additional wells were installed at the request of IEPA—two in the dolomite and one in the till. In 1984, IEPA installed seven wells in the glacial drift in order to clarify interpretations of the site geology. The first extensive investigation of the surficial geologic deposits at this site was conducted in conjunction with this drilling program.

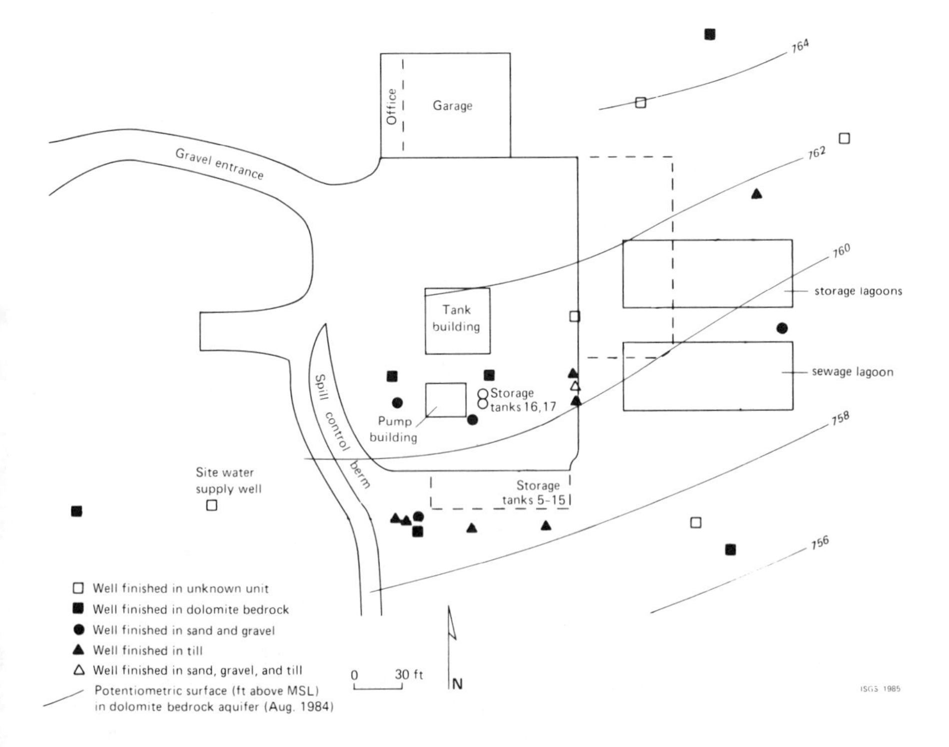

Figure 5. Map of Site 10 showing storage tank locations, approximate location of storage and sewage lagoons, well locations, and potentiometric surface in dolomite bedrock aquifer.

The current groundwater monitoring system is considered inadequate, primarily due to the lack of reliable monitoring wells completed in the bedrock. For the purpose of detecting groundwater contamination, the value of the monitoring wells installed prior to 1984 is questionable since the construction records for these wells do not exist, and the logs describing the geologic materials are vague and confusing. Therefore, the information gathered from these wells is not reliable. In addition, since no wells were installed in the dolomite after 1983, there are no reliable monitoring wells completed in the dolomite aquifer. The hydrogeology of this site is too complex to describe based on data obtained from the wells completed in the glacial drift; wells completed in the dolomite are also necessary. Thus, additional wells are needed in the dolomite, especially downgradient of the waste lagoons. Also, data from wells completed in the dolomite are necessary to define the regional groundwater flow, vertical groundwater gradients, and the position of the contaminant plume.

In this case history, the importance of a complete and accurate understanding of hydrogeologic conditions at a site in the design of a groundwater monitoring system is demonstrated. The groundwater monitoring system for this site was apparently designed with little regard for the site hydrogeology. The case history revealed the potential ramifications of such a design--contamination of a domestic groundwater supply.

This case history also demonstrates the need to have an experienced and qualified professional in charge of the drilling operation, who can carefully and thoroughly describe geologic samples. Also, a need exists for well construction records to be complete and properly filed for future reference. Without this information, the monitoring wells will lack scientific value, resulting in the need to conduct another hydrogeologic investigation of the site.

SUMMARY AND DISCUSSION

The results of the evaluation of the groundwater monitoring systems for ten hazardous waste management facilities, including the three case histories discussed in this paper, have shown the importance of an accurate and complete hydrogeologic site description to the design of a groundwater monitoring system. Based on these evaluations, a list of hydrogeologic considerations which were incorporated into or neglected in the design of groundwater monitoring systems was compiled. Table 1 lists these factors which should be considered by designers of groundwater monitoring systems as they evaluate the description of the site hydrogeology for accuracy and completeness or plan a site hydrogeologic investigation. All of the factors listed in Table 1 may not apply to all sites, but many of these items will apply to every site. Consideration of these points during the design of a groundwater monitoring system should lead to the design of more effective groundwater monitoring systems and save the operator of a hazardous waste management facility the expense of conducting a supplemental hydrogeologic site evaluation.

Also, designers of groundwater monitoring systems might benefit by adopting a new objective when designing these systems. Instead of designing a groundwater monitoring system to merely fulfill the requirements of the federal and state regulations, the groundwater monitoring system should be thought of as an operator's defense against the potentially high cost of liability resulting from groundwater contamination. An effective groundwater monitoring system should detect any contamination before the problem significantly threatens the environment. In order to design an effective groundwater monitoring system, the site hydrogeology must be fully understood.

Table 1. Hydrogeologic Considerations in the Design of Groundwater Monitoring Systems.

Regional Geology

Has the regional geology been investigated?

Can the site geology be approximated from knowledge of regional geology?

Have the field investigations been planned based on regional and near-site geologic conditions?

Field Investigations

Have a sufficient number of borings been drilled to characterize the site geology?

Have samples been continuously collected by a qualified professional? Have these samples been described, at least briefly, on-site by a qualified professional?

Have samples been returned to a laboratory to determine stratigraphic relationships and perform characterization tests?

Have hydraulic conductivity tests been run on appropriate formations? Were field or lab hydraulic conductivity tests conducted? Were these tests run properly?

Are the well construction records and boring descriptions complete and accurate?

Does the data collected correlate with the regional geology? (This should be checked at all stages of site characterization.)

Site Hydrogeology

Have all the aquifers and/or zones of contaminant transport (including sand lenses and fracture zones) beneath the site been identified?

Were the piezometers/monitoring wells properly installed such that the data collected are accurate and complete?

Have a sufficient number of piezometers/monitoring wells been installed to determine:

The potentiometric surface for each aquifer/zone?
The direction of groundwater flow for each aquifer/zone?
The vertical gradient between aquifers/zones?

Have the data been collected over a sufficient period of time to consider seasonal fluctuations of the potentiometric surface, direction of ground-water flow and vertical gradients?

Have the monitoring wells been installed sufficiently in advance of site operation to allow for an accurate assessment of the initial ground-water quality?

REFERENCES

1. Everett, L. G. Groundwater Monitoring: (Schenectady, NY: General Electric Company, 1980), p.440.

2. Nacht, S. J. "Ground-water Monitoring System Considerations," Ground Water Monitoring Review, 3(2):33-39 (1983).

3. Graves, L. S. "Ground-water Monitoring Requirements of RCRA," Ground Water Monitoring Review, 1(2):34-36 (1981).

4. Herzog, B. L., B. R. Hensel, E. Mehnert, J. R. Miller, and T. M. Johnson, "Evaluation of Groundwater Monitoring Programs at Hazardous Waste Disposal Facilities in Illinois," Illinois State Geological Survey, in press, p. 190, (1986).

INVESTIGATING SUBSURFACE FUEL RELEASES

Tim Holbrook, ERT, A Resource Engineering Company

ABSTRACT

Investigations of subsurface fuel releases in an urban setting are complicated by a variety of factors. A preliminary investigation is necessary to review any and all existing information available for a given site prior to formulating plans for soil or groundwater sampling. Most critical at this stage is the identification of buried water, sewer, or other utility lines that form the "urban hydrogeology" of the site.

The next recommended step is to plan a soil vapor survey to define the occurrence of fuel vapors in the vadose zone. Methods for obtaining specific and relative levels of contamination at several points are discussed. A protocol for ambient temperature headspace (ATH) analysis is presented. Case studies in which this method have been used are discussed. Based on the results of a soil vapor survey, the design, location, and number of monitoring and recovery wells can be optimized.

INTRODUCTION

Investigators, operators, owners, and managers of underground storage tank (UST) sites need an economical way to determine the presence and extent of subsurface fuel contamination. Information gained as a result of a preliminary investigation and a soil vapor survey facilitate construction of a reliable information base. This allows informed and economical decisions about the need, number, and placement of monitoring wells to be made.

Geotechnical and Geohydrological Aspects of Waste Management, D. J. A. van Zyl et al., Eds., © 1987 Lewis Publishers, Inc., Chelsea, Michigan – Printed in USA.

Primary goals of a subsurface investigation could be one or more of the following:

1) To determine the existence of any contamination if pipe or tank integrity test results are inconclusive.

2) To fulfill a requirement of regulatory agencies.

3) Gather information to evaluate the feasibility of fuel recovery or other remedial actions.

4) Produce information critical to the appropriate number, the location and the design of monitoring and recovery wells.

To set the stage, an investigation might be triggered by one or more of the following events: fumes in a nearby basement, tank system integrity test failure, increasing inventory losses, or nearby well contamination. Other less immediate reasons for investigation may include an environmental audit, or an insurance requirement.

PRELIMINARY INVESTIGATION

The first step in a subsurface investigation is a preliminary investigation in which existing information is used to: 1) estimate the leak volume and/or duration, 2) identify migration pathways, 3) estimate the extent of contamination, 4) locate the receptors, and 5) locate all possible sources. Sources of existing information include the facility operator, regulatory agencies, and utilities companies.

Operator Information

Information that could be available from the operator is as follows: 1) inventory records, 2) the method of inventory control, 3) results of tank and line integrity tests, 4) fuel storage system repair records, 5) original installation drawings, 6) corrosion protection system existence and operation, 7) site history, and 8) surrounding land use.

Regulatory Agencies

Regulatory agencies, may be able to provide the following information: (enforcement status may limit availability) 1) data from a nearby investigation, 2) site geology, 3) groundwater flow directions, 4) permitted well locations, 5) spill or leak reports, and 5) highway department aerial photographs, designs of drainage networks, cut and fill operations, and geotechnical studies. Local USDA soil conservation service offices may have soil surveys that are useful in defining the near surface geology.

Definition of the "urban hydrogeology" during a preliminary investigation requires location of all of the possible permeable conduits (composed of porous backfill materials) that are formed in providing utilities to the site. City and County engineering departments may provide details of these features including location, depth, dimensions, backfill composition, slope, and access points. This information will assist in directing a drilling program to detect fuel in these migration pathways.

Utility Companies

Additional features of the urban hydrogeology may be located with the assistance of utility companies. Electric, gas, water, television and telephone cables are susceptible to influences from hydrocarbons and solvents. The location, depth, and the construction of these utility lines can be critical. Not only the backfill materials but the slope of the trench and the surrounding soil type are important. If a utility company has excavated in the area lately, observations from workers or supervisors on the site might be available.

INITIAL SITE INVESTIGATION

The goal of the initial site investigation is to obtain information necessary for planning a soil vapor survey. Facility related information to gather on-site includes: Tank system equipment details, the presence of any leak detection equipment, observations that the staff may have had during fuel system repairs, inventory records, leak history, and wastes (antifreeze, solvents, degreasers) generated, stored, and disposed of. The extent of surface materials (concrete, asphalt, vegetation) should be sketched in a field notebook at this time. By removing the fuel dispenser cover, leaks in the dispenser itself can be detected. Hydrocarbon vapor testing of nearby manholes located in the preliminary off-site investigation can be accomplished with a combustible gas indicator and a photoionization detector.

SOIL VAPOR SURVEYS

A soil vapor survey is the measurement of relative or specific volatile hydrocarbon concentrations in soil pores in the unsaturated zone at various points, distributed vertically and horizontally. This is not a groundwater investigation, this is a vadose zone, or unsaturated zone investigation. After sufficient site information has been accumulated, a soil vapor survey can be designed that concentrates on possible zones of hydrocarbon migration.

In the unsaturated zone, hydrocarbons, can exist in the vapor phase in soil pores, they can be sorbed on soil particles, and they can be free liquid in soil pores. Over a sufficient period of time, an equilibrium is established between the vapor, sorbed, and liquid hydrocarbon phases. Concentrations of volatile hydrocarbons in the vapor phase in soil pores are generally proportional to the distance from the source of contamination. In other words, the closer a soil pore is to the fuel leak source, the higher its hydrocarbon vapor concentration. By obtaining soil vapor data at vertically and horizontally distributed points and plotting the data, the extent of subsurface contamination can be defined [1].

Having a good idea of where the vadose zone is contaminated allows optimization of the placement, and minimization of the number of monitoring wells. Depending on the soil vapor survey method used, the occurrence of hydrocarbons in recoverable quantities can be evaluated. The placement of hydrocarbon recovery wells can also be optimized. Therefore, a soil vapor survey can provide, valuable information at an UST site.

Soil Vapor Survey Methods

Methods currently in use include surface and subsurface flux chambers, driven ground probes, buried static accumulators, and soil sampling with various forms of sample head space analysis. These methods, described below, generally employ gas chromatography to yield specific concentrations of volatile hydrocarbons. At sites contaminated with strictly gasoline or other well characterized constituents, the sophistication of gas chromatography is generally not needed to determine relative contamination levels. This prompted the development of the ambient temperature headspace (ATH) method. This method yields relative levels of hydrocarbon concentration. ATH results are usually correlated with specific methods by analyzing a subset of samples with a gas chromatograph.

Surface Flux Chamber

A surface flux chamber is a small plastic hemisphere that is placed over the surface of the ground and sealed to the ground surface. Impermeable surface layers have to be removed first. Soil vapors are extracted from the enclosed hemisphere by a gas sampling pump and fed to a gas chromatograph. One advantage of this method is that it can be used in very rocky soil types. Another advantage is that contaminant charaterization is possible with the use of a portable gas chromatograph. Disadvantages of the surface flux chamber are the influence of dilution from the vapor as it enters the chamber from surrounding soil and the influence of the surface temperature.

Subsurface Flux Chamber and Subsurface Probes

A subsurface flux chamber is similar to the surface flux chamber, only it is placed in a augered hole at incremental depths. It is an inverted can connected to a gas sampling pump that pulls vapors out of the soil matrix and into a gas chromatograph. Placing this instrument at incremental depths can help characterize the vertical extent of contaminants.

A subsurface probe is a rigid tube with a removable tip. It is driven down into the ground at certain sampling locations. The sleeve that is used to drive the probe is then retracted to form a space between the sleeve and the point. This forms a space were vapors can enter and be drawn to a gas chromatograph by a sampling pump.

Some of the advantages of the subsurface flux chamber and the subsurface probe are 1) no soil sampling, 2) restriction of surface influences, and 3) variable depths can be reached depending on the length of the tube and the site geology. Contaminants can be characterized if that is a concern. Disadvantages include problems with gravelly soils and it is time consuming.

Accumulation Devices

Accumulation devices are typically charcoal coated wires contained in small tubes which are buried and exposed to the soil atmosphere. After waiting for 30 to 60 days the tubes are dug and analyzed for contaminants of concern. Contaminants are desorbed by heating the coated wires in the laboratory and injecting vapors into a gas chromatograph. The advantages of accumulation devices include stability of samples during transport [2] and accurate characterization of the contaminants [3]. Disadvantages are 1) finding the probes after burial may be difficult, 2) a very small portion of the soil matrix is sampled depending on the time probes are left in the ground, and 3) the results are not available for 40 to 70 days.

Soil Sampling and Sample Headspace Analysis

This method consists of collecting discrete or composite soil samples. Sample containers are partially filled with soil to form a vacant headspace in the glass container. The headspace gas in each sample is then analyzed for organic vapors following EPA [4] or other methods [5].

A tool commonly used to gather soil samples at depth is a split spoon sampler. After augering to the target sampling depth, the split spoon is connected to a solid steel rod and then driven into the ground with a drop hammer. As the tool is driven, the tube fills up with the soil sample. After driving to the desired sampling depth, the tube is extracted. The halves of the sampler split longitudinally when retainer rings are unscrewed. The soil can then be extracted from the sampler and placed in a container and sealed. The advantage of using this tool is that it is easy to decontaminate and it minimizes vertical cross contamination since only the center of the sampler is in contact with the soil sample. Unfortunately, the spilt spoon method is very time consuming.

Soil samples can also be obtained using a hydraulically driven, solid-stem auger. Surface sod, asphalt, or concrete is removed prior to drilling. A decontaminated 5 cm diameter, 1.5 meters long auger is then advanced in 1.5 meters increments. No samples are collected from the top 0.75 meter due to the possible influence of surface contamination from sources unrelated to a subsurface release.

Soil samples are composited over 0.75 meter (30-inch) intervals directly from the auger flights. Samples are placed in clean glass sample containers. A new, disposable, surgical glove is used to handle each sample. The sample containers are filled half full, to create a head space, and then sealed with a double layer of aluminum foil and a jar ring and lid.

At the completion of each 1.5 meter interval, a portable photoionization detector (PID) is used to indicate the presence of hydrocarbons in the borehole and on the auger as it is extracted. If the PID detects hydrocarbons, the borehole is reamed to a 10 cm diameter before proceeding with the next 1.5 meter interval to avoid vertical cross contamination. A decontaminated 5 cm auger is then used for the next 1.5 meter interval. Augering continues in increments of

1.5 meters until saturated conditions compromise sample integrity. Boreholes are logged using the Unified Soil Classification System.

All buried utility cables and pipes near each borehole are located prior to drilling. Drilling begins as near to the underground tank backfill as possible. If contamination is detected at the top of the borehole by the PID, the next borehole is drilled approximately 6 meters feet away from the tank. This continues in four directions along roughly perpendicular axes. Drilling continues outward along an axis until the PID does not detect hydrocarbons on auger flights or in the borehole. Headspace and analysis results are utilized on a daily basis to confirm if the lateral and vertical extent of contamination has been reached.

Sampling equipment must be decontaminated between each sample run. Decontamination procedures include a surfactant wash, deionized (DI) water rinse, methanol rinse, and a final DI rinse. Equipment is dried before use and checked with the PID to assure proper decontamination.

Sampling equipment rinsate samples should be collected as a quality control procedure. Deionized water is sprayed onto the sampler or auger flights and collected in a sample jar, sealed, and analyzed in procedures identical to soil samples.

Ambient Temperature Headspace (ATH) Protocol

Headspace analyses are performed daily on samples collected the previous day. Samples are stored in a constant temperature environment (20°C) for a minimum of 12 hours. This allows all samples from the previous day's sampling to reach the same temperature prior to analysis. This also ensures that hydrocarbons uniformly volatilize from the soil into the container head space. Filling the glass sample containers half full during sampling creates a vacant headspace of approximately 300 ml over a soil sample of approximately 300 grams. This vacant headspace is sampled by removing the steel lid and retainer ring and piercing the double aluminum foil seal with a probe extension connected to photoionization detector as shown in Figure 1.

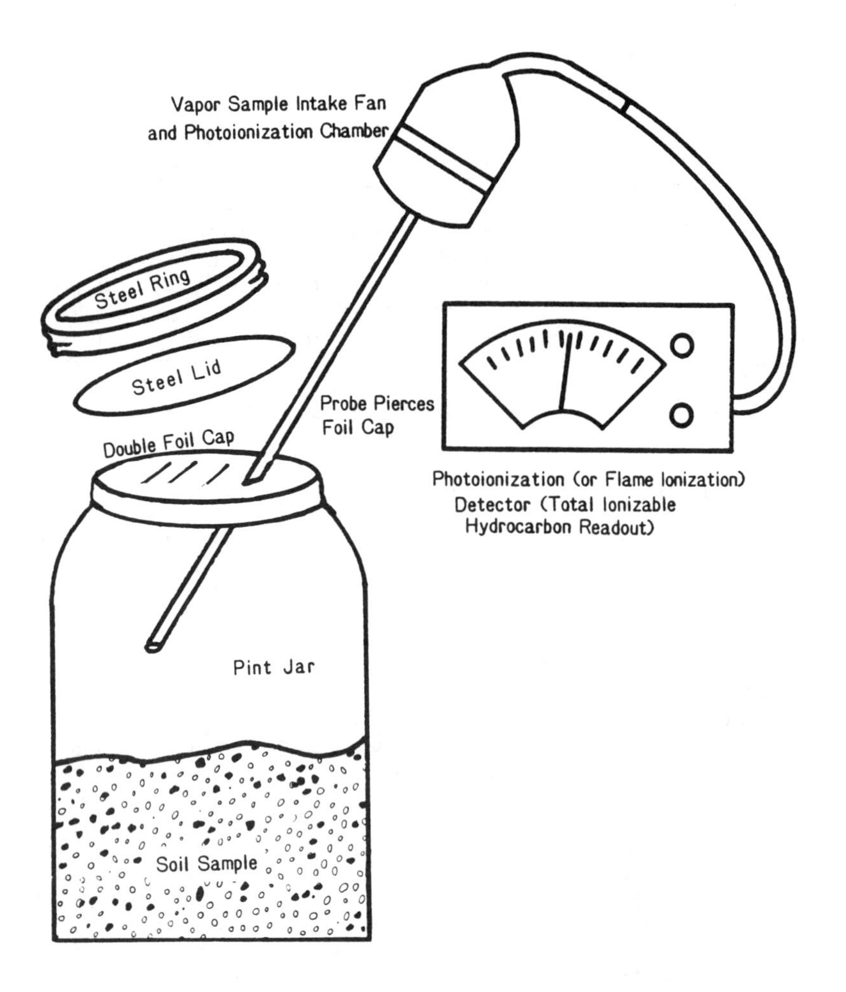

Figure 1. Ambient temperature headspace analysis.

One PID, manufactured by HNU Systems Inc., has a built in fan that draws vapors into the ionization chamber at a rate of 100 ml per minute. At this rate, it would take approximately 3 minutes or 180 seconds to completely evacuate the headspace in a sample container. The HNU reaches a peak response within 5 seconds. Therefore, the headspace reading should be taken between 5 and 10 seconds after piercing the foil seal. This is long enough for the meter to respond, but far short of withdrawing enough vapor to affect the vapor equilibrium in the sample container. Resulting HNU readings are in ppm of total ionizable hydrocarbon based on a benzene standard.

MONITORING WELLS

Soil vapor survey information collected at incremental depths can be used to decide if monitoring wells are necessary. If the data determines that the vertical extent of contamination is far above the water table, monitoring wells may not be necessary. A high concentration of hydrocarbon vapors may indicate the presence of a free fuel layer. The center of such a concentration is an efficient location for a monitoring or recovery well. The well is then in position to accumulate fuel from saturated soil pores. The rate of fuel accumulation in the well can be used to evaluate the feasibility of fuel recovery. The shape of a fuel plume, determined with relative or specific soil vapor survey data, can be used to locate a downgradient detection well and an upgradient, background well.

CASE STUDIES

A hydrocarbon seep on a stream bank in northern Wyoming was discovered to be discharging gasoline into a major river. Three possible sources of leaking gasoline were identified and are shown in Figure 1. Site A, was within 50 meters of the seep and had 5 underground storage tanks. Site B was approximately 125 meters away from the seep and had 3 USTs. The third possible source Site C was more than 160 meters away. A preliminary investigation revealed that the three possible sources were all within 10 meters of a buried storm sewer leading to the river near the hydrocarbon seep. The storm sewer trench had been backfilled with coarse sands and gravels. The surrounding geology was predominantly clay. A decision was made to conduct a soil vapor survey, using soil sampling and ATH analyses to determine which site(s) could be responsible for the hydrocarbon seep.

A series of 21 shallow boreholes were completed to approximately 4.5 meters and soil samples were collected at 0.75 meter intervals for ATH analyses. Results shown in Figure 2 demonstrated that Site C could not have been responsible for the seep since all samples from the four boreholes to the north of Site C hand no detectable hydrocarbon vapors. Headspace analyses of samples taken near Site B showed that the nearby storm sewer could have been responsible for transporting free hydrocarbons to the seep area at the river. This was further evidenced by elevated hydrocarbon concentrations in soil samples from boreholes BH-5, BH-18,

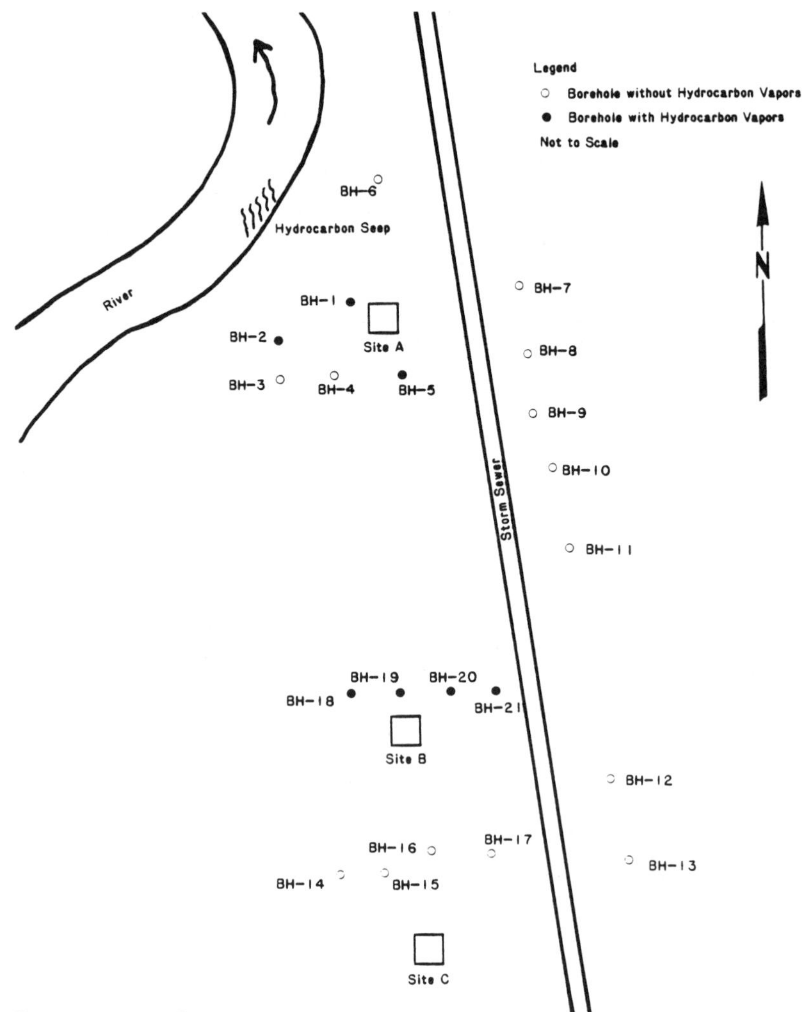

Figure 2. Case history.

BH-19, BH-20, and BH-21. Boreholes drilled at Site A produced soil samples with elevated hydrocarbon vapors by the ATH method. Data from this soil vapor survey was used to focus further investigative efforts on Site A and Site B. Additional costs of investigating Site C with monitoring wells were saved.

In Denver, Colorado, a soil sampling and ATH soil vapor survey was conducted to determine the extent of a surface spill of methylene chloride and perchloroethylene. A series of 60 soil samples were gathered at incremental depths for ATH analysis. Results indicated all contamination was found in the unsaturated zone. Therefore, monitoring wells were not required. ATH analyses were also used to direct excavation of contaminated soil. ATH results were found to be well correlated with laboratory gas chromatograph analyses of soil samples.

Owners of a manufacturing plant in northern Colorado were not satisfied with the accuracy of underground tank integrity test methods. Therefore, they opted to conduct soil sampling with ATH analysis to detect the presence of leaking gasoline and solvents from four UST systems. Hydrocarbon contamination was detected near three USTs. The goal of the investigation then changed from detection to definition of the extent of contamination.

By using the ATH method on over 200 soil samples obtained at incremental depths, the extent of contamination in the unsaturated zone was defined. Geologic and ATH data supported that all contamination was confined to the immediate vicinity of the tank installations. At each of the three leaking UST system locations, the ATH data was used to place one monitoring well in the center of the unsaturated zone hydrocarbon plume and one just outside and downgradient from the plume. The well screened in the plume was used to evaluate hydrocarbon recovery feasibility.

CONCLUSION

The use of thorough preliminary investigations and soil vapor surveys can minimize the number and optimize the location of expensive monitoring wells. The information derived from the exercises can also optimize the placement of fuel recovery trenches and/or wells. Tank and piping integrity tests can detect major leaks but are not always effective in finding small leaks. The ultimate tank test is a soil vapor survey. Soil sampling and ATH analysis is a cost effective soil vapor survey method. Advantages of soil sampling and ATH method are: 1) generation of geologic data, 2) rapid availability of results, and 3) economical analytical procedures. This method provides relative hydrocarbon concentrations that can be used to detect the presence of hydrocarbons above background levels and to delineate volatile contaminant plumes. Results of a research project designed to correlate ATH results with established VOA methods will soon be available.

REFERENCES

1. American Petroleum Institute, Detection of Hydrocarbons in Groundwater by Analysis of Shallow Soil Gas Vapor, Publication Number 4394, API, Washington, D.C. 1985.

2. Kerfoot, H. B., and C. L. Mayer, "The Use of Industrial Hygiene Samplers for Soil-Gas Surveying" Groundwater Monitoring Review, Fall, 1986, pp. 74-78.

3. Voorhees, K. J., J. C. Hickey, and R. W. Klusman, "Analysis of Groundwater Contamination by a New Surface Static Trapping/Mass Spectrometry Technique", Analytical Chemist, Vol. 56, No. 13, November 1984.

4. U.S. Environmental Protection Agency.

5. Ioffe, B. V. and U. Vitenberg "Head Space Analysis and Related Methods in Gas Chromatography", Wiley Interscience, New York, 1984.

HYDROGEOLOGIC ASPECTS OF RECLAMATION OF THE RAY POINT TAILINGS FACILITY: A CASE STUDY

Louis L. Miller and Lyle A. Davis,
Water, Waste and Land, Inc., Ft. Collins, Colorado

INTRODUCTION

In the past few decades, disposal of wastes from uranium mills has been accomplished by transporting the waste material in slurry form to impoundments constructed specifically to retain the wastes. Because of the decline in demand for uranium products, many of the uranium mills in the country have been decommissioned making it necessary to initiate reclamation of the tailings facilities. Recent concern over pollution of groundwater resources has resulted in regulations which requires a tailings reclamation plan to address potential impacts to groundwater as a result of seepage from the reclaimed tailings impoundment. The purpose of this paper is to present a case study of the hydrogeologic aspects of reclamation of a tailings impoundment located in south Texas.

SITE DESCRIPTION

The tailings impoundment is located in Live Oak County, Texas, approximately 120 kilometers (75 miles) south of San Antonio. The climate of the area is generally mild with an average annual temperature of 22 C (72° F) and a mean annual precipitation of about 775 mm (30.5 inches). Because of the warm temperatures, snow rarely occurs and the precipitation consists almost entirely of rainfall. In a normal year, most of the precipitation occurs in the summer and early fall with September being the wettest month of the year.

Geotechnical and Geohydrological Aspects of Waste Management, D. J. A. van Zyl et al., Eds., © 1987 Lewis Publishers, Inc., Chelsea, Michigan—Printed in USA.

Although substantial rainfall is received in the area, potential evapotranspiration is also high with an average pan evaporation, as measured at sites near the impoundment, of about 1970 mm (77.5 inches).

The land surface in the vicinity of the tailings impoundment is generally level and moderately dissected. Broad, gently rolling hills and level uplands are typical of the area and local relief is usually on the order of 10 to 20 meters/kilometer (50-100 feet/mile). Soils in the area range from silty loam to clay and are used primarily as cropland, pasture, and rangeland.

The tailings basin was constructed in 1970 and tailings disposal was initiated in October of that same year. The impoundment was utilized until March of 1973 when the mill was temporarily shut down. During operations, approximately 368,120 cubic meters (481,480 cubic yards) of tailings materials were placed in the impoundment. Because of low uranium prices, the operators decided to decommission the mill and the initial reclamation plan was submitted to the regulatory agencies for review in May, 1982. Because of concerns expressed by the regulatory agencies and changes in regulations, the initial plan was modified somewhat and the revised plan was submitted for review in April, 1985. The tailings disposal basin has an area of approximately 16.2 hectares (40 acres). Figure 1 depicts the tailings basin prior to initiation of reclamation activities.

Geologic Setting

The site is located in the South Texas Gulf Coastal Plain within the Rio Grande Embayment of the Gulf Coast Geosyncline. The flank of this basin can be characterized as a series of clastic sedimentary strata of Tertiary and younger age. The beds dip to the southeast and east at approximately 1°. The strike of the formations is generally parallel to the coast. The depositional environment varies throughout the region from high velocity fluvial and deltaic conditions to marginal marine and lagoonal conditions. In general the thickness of the beds increases toward the coast.

In the vicinity of the site, the Eocene Jackson Group is the lowest formation of interest. In general this formation consists of fine grained sands, silts and clays of pyroclastic origin that were deposited in a marine or barrier beach environment. Beds of lignite are common in the Jackson Group and are often intercalated with siltstones and sands. The clays which occur in the Jackson Group are primarily bentonitic and are alteration products of degraded tuffaceous material. The Catahoula Formation occurs

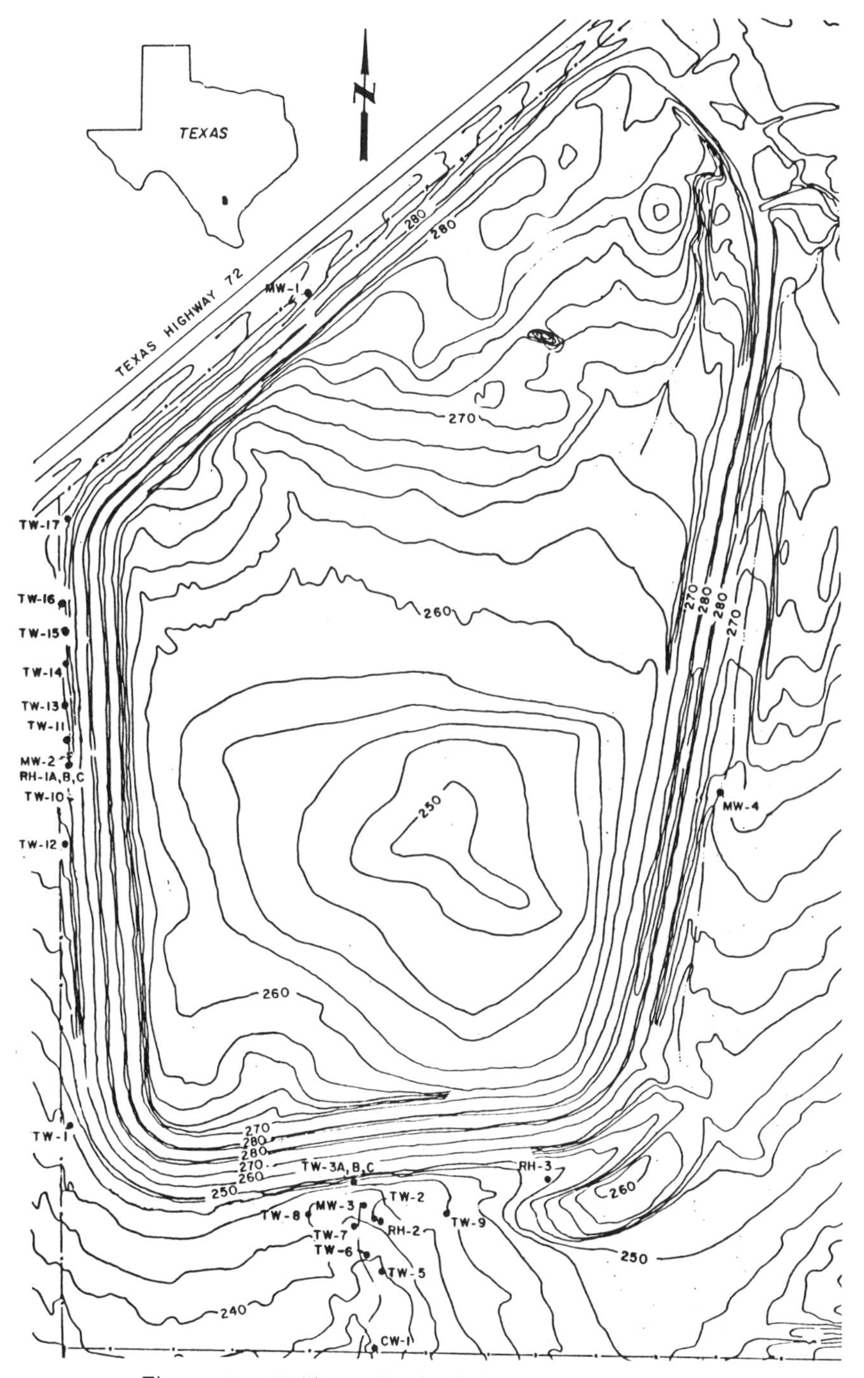

Figure 1. Tailings Basin Prior to Reclamation

above the Jackson Group and consists of a massive thickness of up to 275 meters (900 feet) of bentonitic clays and silty clays. This formation outcrops in the immediate vicinity of the tailings impoundment. The Oakville Sandstone Formation lies on top of the Catahoula to the east of the impoundment. This unit typically consists of coarse grained, arkosic sands overlain by siltstones and dense blue clays. The Oakville was the primary uranium producing formation in the vicinity of the site. This unit outcrops to the east of the tailings impoundment.

Hydrologic Setting

The impoundment was constructed at the head of a small tributary to Sulfur Creek which ultimately empties into the Nueces River. Although Sulfur Creek is perennial in nature, the small tributaries in the vicinity of the tailings impoundment experience flow only during and after precipitation events. Because of the location of the impoundment near the upper reaches of the drainage basin, diversion of surface runoff was easily accomplished during construction.

Groundwater recharge in the vicinity of the tailings impoundment is limited because of the tight nature of the surficial soils and that annual potential evapotranspiration greatly exceeds annual rainfall. Groundwater usage in the area is generally limited to the Oakville formation which is extremely thin and unsaturated in the immediate vicinity of the impoundment. Most producing wells in this formation occur to the south and east of the site. The Catahoula formation does not represent an aquifer in the immediate site area although limited water may be locally available in silt and sand lenses.

RECLAMATION PLAN

The final reclamation plan developed for the site was designed to produce a reclaimed surface that provides long-term stability, controls the emanation of radon within acceptable limits, and provides long-term containment of radioactive and non-radioactive tailings constituents without long term maintenance. The plan selected consists of a surface configuration which controls all surface runoff from the site and discharges to the southeast. The final configuration of the reclaimed tailings impoundment is shown on Figure 2. This design offers several advantages over other alternatives considered during selection of a final reclamation plan. First, the gentle slope to the east and southeast provides an ideal route to discharge

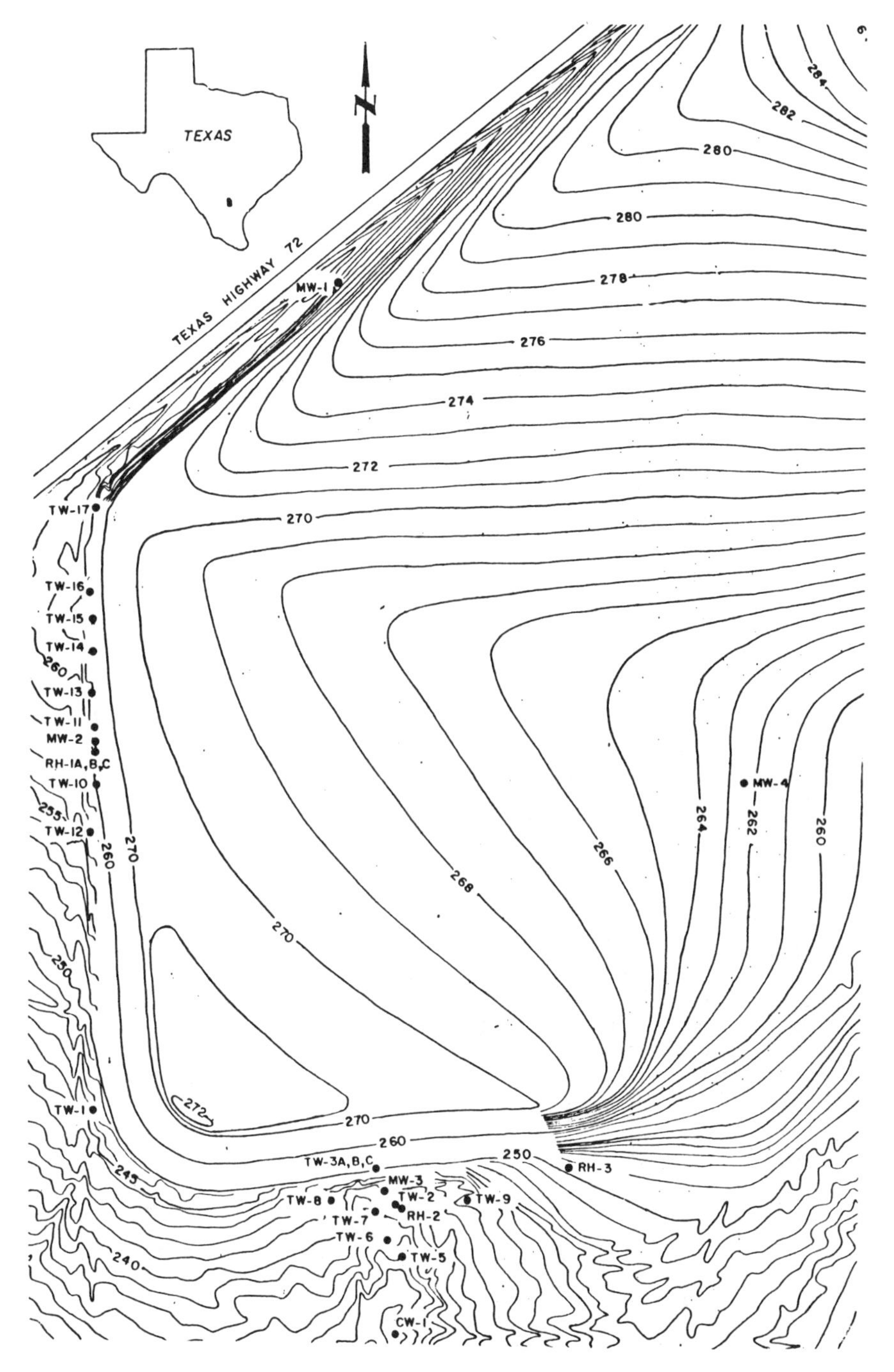

Figure 2. Reclaimed Tailings Basin

runoff from the site in a designed swale without causing erosive velocities even under extreme events. Second, the majority of the runoff is diverted away from the 5H:1V reclaimed and shortened embankment slopes via the designed swale thereby increasing the long-term stability of those embankments. Third, by controlling the surface runoff in a designated area, accurate estimation of erosion potential was possible and the swale was designed to preclude long-term erosion and gully formation.

The reclamation plan was designed to divert water from the surface of the reclaimed tailings impoundment thereby limiting the amount of infiltration into the covered tailings. In addition, approximately 1.2 meters (4 feet) of cover material are required to insure that radon attenuation is accomplished. The cover material selected is a highly plastic clayey sand which will also serve to further limit infiltration.

GROUNDWATER IMPACTS AND MOISTURE MIGRATION

Because the tailings impoundment was constructed on the extremely thick Catahoula formation, seepage from the facility during operations was assumed to be negligible. In addition, the final reclamation plan was developed to preclude additional infiltration into the tailings material. Based on these considerations, it would seem that groundwater impacts would be minimal at the site. Unfortunately, resistivity studies conducted by the Texas Department of Health (TDH) indicated that moisture migration was occurring in excess of predicted amounts. In addition, water quality samples collected from shallow wells constructed immediately outside the impoundment indicated that moisture had indeed migrated further from the impoundment than would be expected given the tight nature of the clays at the site. It therefore became necessary to determine the extent of the moisture migration and to evaluate the future impacts to groundwater resources in the vicinity of the impoundment.

Well Installation

At the present time 30 wells exist in the immediate vicinity of the tailings impoundment. The first four wells were installed in 1980 and are approximately 15 feet deep. Additional wells were installed in 1984, 1985 and 1986. A complete listing of well completion details is provided in Table 1 and well locations are shown on Figures 1 and 2.

Table 1. Well Completion Details

Well ID		Installation Date	Measuring Point Elevation	Ground Surface Elevation	Well Bottom Elevation
MW-1		1980	267.41	267.41	251.00
MW-2		1980	262.90	260.36	245.57
MW-3		1980	241.10	239.14	223.95
MW-4		1980	267.70	263.75	249.95
TW-1		03/11/86	247.34	245.88	199.65
TW-2	*	03/12/86	238.73	238.73	198.73
TW-3		03/12/86	244.59	243.06	215.29
TW-3A		03/15/86	244.27	242.37	204.27
TW-3B		03/16/86	243.54	242.24	213.67
TW-4	*	03/12/86	241.11	241.11	237.11
TW-5		03/12/86	237.52	236.02	198.22
TW-6		03/13/86	238.20	236.70	198.45
TW-7	*	03/13/86	237.98	237.98	197.98
TW-8		03/13/86	243.44	240.94	200.94
TW-9		03/13/86	244.05	242.95	204.95
TW-10		03/14/86	260.63	259.53	220.68
TW-11		03/14/86	262.26	260.46	228.46
TW-12		03/21/86	258.10	256.75	218.10
TW-13		03/22/86	262.44	261.56	222.44
TW-14		03/26/86	264.16	262.01	224.16
TW-15		03/26/86	263.75	262.30	223.75
TW-16		03/26/86	264.63	263.31	224.63
TW-17		03/26/86	263.55	261.77	223.55
CW-1		03/15/86	236.18	233.93	187.43
RH-1A		11/21/85	261.12	259.52	221.12
RH-1B		03/15/86	260.74	258.99	219.99
RH-1C		03/15/86	260.76	258.76	230.26
RH-2		11/20/85	240.80	239.20	198.90
RH-3		12/15/84	250.20	248.20	240.60
RH-3A		11/20/85	250.09	248.39	209.09
WWL-3		12/15/84	275.83	274.33	255.83
WWL-2		12/15/84	267.10	265.31	244.10
WWL-1		12/15/84	274.25	271.33	213.75

* indicates well was not completed.

All wells with the exception of WWL-1 and WWL-3 were completed in the Catahoula formation or soils overlying the Catahoula. Well WWL-3 was completed in tailings materials to allow tailings water levels to be monitored. WWL-1 was is about 60 feet deep and is completed in the Oakville formation. It is not shown on Figures 1 and 2 since it is some 2,000 feet due east of MW-4. The wells were concentrated to the south and west of the tailings impoundment since resistivity data made available by the TDH indicated that moisture migration had occurred primarily in these two areas. Soil and rock samples were collected during the drilling of all wells with the exception of the MW series of wells.

Because resistivity survey data and water levels in the MW wells indicated that moisture existed at depths of 3 to 10 meters (10 to 30 feet) in areas where the wells were being placed, extreme care was taken in drilling and sampling the RH and TW wells. For most of the holes, continuous sampling was attempted but due to the extremely tight nature of the soils, this was not always possible. Based on inspection of samples and the drilling difficulties encountered, it was anticipated that free water would not be found in any of the holes.

RH-1A and RH-2 were cased immediately after drilling. Free water was found in both of these wells after about 24 hours. The remainder of the RH and all of the TW wells were left uncased for several days. Periodic inspection of these open holes was made in an attempt to determine the level at which water was entering the hole. Of the holes that eventually made water, it took from a few hours (TW-3) to nine days (TW-5) for water to appear. In all cases, it was not evident that water was seeping from a unique layer.

Visual inspection of the sampled materials indicated the soil below a depth of 3 meters was a very stiff, moist, red to green, silty clay. The surficial soil consisted of a stiff, slightly moist, calcareous silty sandy clay. No apparent visual differences could be detected below a depth of about 3 meters.

Laboratory Testing

Upon close visual inspection of the retrieved samples, some subtle differences in the samples could be detected. Some siltier layers in some of the core samples, especially that from TW-3 were identified. Laboratory testing including Atterberg limits, density and moisture content, hydrometer analysis and permeability were performed on a large number of samples.

Atterberg limit tests indicated the material below a depth of about three meters (10 feet) classified as either a ML or MH type soil. The calculated degree of saturation correlated well with the water level readings. Hydrometer analyses were the best indicator of the subtle visual differences. Most of the soils encountered below a depth of three meters (10 feet) were comprised of between 40 to 60 percent silt size particles and from 30 to 50 percent clay sized particles with the remainder, 0-15 percent, being sand sized particles. The layers of visually siltier materials had particles size distribution of 75 to 80 percent silt and 20 to 25 percent clay sized particles. Hydraulic conductivity tests performed on samples representative of both of the materials however showed no significant difference with measured permeabilities ranging from 7 x 10^{-6} to 3 x 10^{-7} cm/sec for both material types.

Although slight differences in particle size distribution for specific layers of material could be determined, permeability testing showed they did not have significantly different hydraulic properties. Furthermore, attempts to demonstrate a continuous silt layer existed from one hole to another proved futile.

Water Level Measurements

Water levels have been collected on a weekly basis since completion of the latest drilling program in April of 1986. Several of the wells (TW-1, TW-9, TW-16, TW-17, CW-1, and WWL-1) have remained dry during the entire monitoring period. Substantial fluctuation has been noted in wells which are located to the south of the impoundment while the wells which are located to the west have remained fairly stable.

Water level elevations as a function of time for wells TW-5, TW-6, and TW-8 are shown on Figure 3. As stated previously well TW-5 remained dry for about nine days after it was drilled on March 12, 1986. Water levels then increased until about mid-April, when they began to decline. By June 18 the water levels had dropped to the bottom of the well. On June 7, a large rain storm of 124.5 mm (4.9 inches) replenished water in the pools within the impoundment. The June 25 water level measurements revealed that water levels were again increasing in TW-5. Water levels continued to rise in the well until mid-July when the trend was reversed. By October 9 the well had again dried up. Water levels in TW-6 have continuously declined since the first measurements were taken in mid-March with the exception of a short period in the latter part of June. After reaching a maximum on June 25, water levels have been dropping continuously. By October 24 the water level had dropped below the bottom of the well. Water levels in well TW-8 have shown less variation than those in the other two wells presumably since TW-8 is closer to the impoundment. Nonetheless, a declining trend is apparent after the well reached a maximum water level on July 11.

Figure 4 depicts water levels as a function of time for wells TW-10, TW-13 and TW-14. As this drawing demonstrates, very little fluctuation in water levels has occurred since the wells reached equilibrium conditions. It is interesting to note that these wells

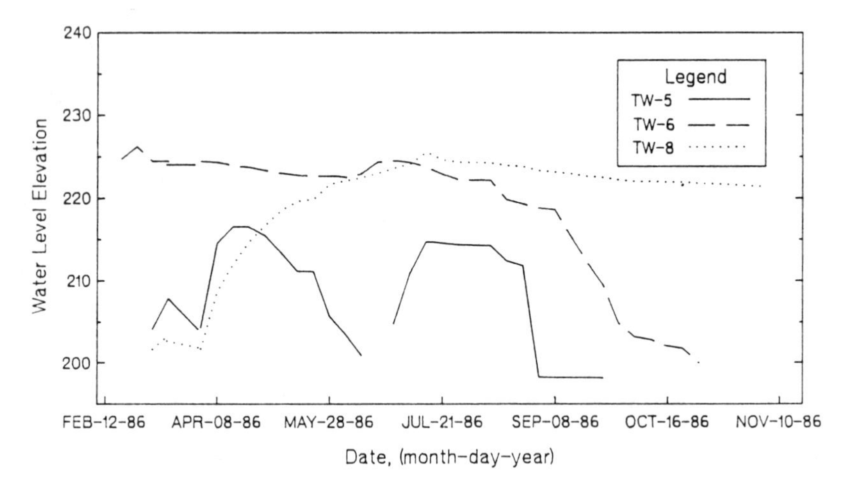

Figure 3. Typical Water Levels South of Tailings Basin.

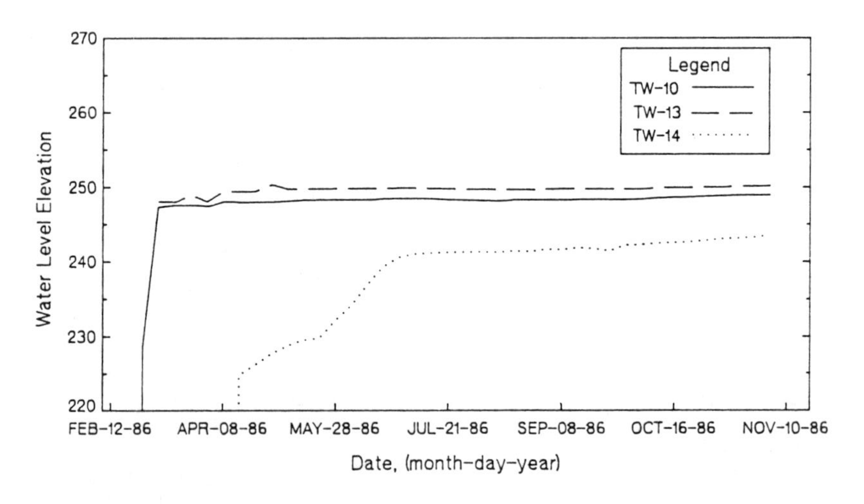

Figure 4. Typical Water Levels West of Tailings Basin.

did not respond to the June 7 rainfall event, at least not to the magnitude experienced by wells located south of the impoundment. However, water levels collected since mid-October indicate that water levels are increasing. Because several rainfall events have occurred since the first of October, the free water pool, which had dried up over the summer months, has returned to the basin. It is likely that this is the cause of increasing water levels to the west of the impoundment, although water levels are continuing to decline to the south of the impoundment.

Water Quality Data

Samples of water from wells MW-1, MW-2, MW-3, and MW-4 have been collected and analyzed on a quarterly basis since 1981. The sample analyses indicate high concentrations of Total Dissolved Solids (TDS), Sulfates (SO_4) and Chlorides (Cl). However, radionuclides and other metal concentrations, with the possible exception of Molybdenum, are near background levels. The pH of the water in these wells has been consistently near neutral indicating that seepage from the impoundment (highly alkaline water existed in the impoundment at the termination of operations) has not reached the wells or that the soils have sufficient buffering capacity to drop the pH and cause most of the metals to come out of solution. It should also be noted that the normal 'tracers' associated with uranium tailings seepage (arsenic and selenium, for example) have not been found in detectable quantities. Further, groundwater in the vicinity is known to have large concentrations of chlorides and sulfates, especially water found in sandy zones within the Catahoula formation.

ANALYSES

As the preceding discussion indicates, some seepage from the impoundment has occurred in the past and minor amounts of seepage continue to occur. Regulations require prediction of the location of contaminants due to seepage from the impoundment at times of 200 years and 1000 years after closure of the facility. Based on the tight nature of the extremely thick Catahoula formation, it would be reasonable to expect very limited contaminant migration in the lateral direction with a majority of any seepage moving vertically downward through the formation. Assuming a vertical permeability of 1×10^{-6} cm/sec, a formation thickness of 200 meters and a gradient of unity, the length of time for a conservative contaminant (e.g. chloride) front to reach the underlying Jackson Formation would be on the order of 630 years. Of course, other contaminants would lag behind the conservative species due to retardation and decay. Clearly, little, if any, impact to groundwater resources should be expected for the next 600 to 1000 years under such conditions.

While the above approach seems valid, field measurements indicate that moisture has migrated a substantial distance (about 150 meters (500 feet)) laterally from the impoundment in the past 16 years. Based on interpretation of water level measurements and resistivity survey data, it is believed that a zone of slightly larger lateral permeability occurs along the western side of the impoundment. This zone is believed to exist at the interface between the Catahoula formation and the overlying soils some of which are derived from both the Catahoula and Oakville formations. It is likely that this zone may outcrop in the northern end of the impoundment, leading to relatively rapid movement of water during periods when water is ponded on the surface of the tailings in the northern portion of the impoundment.

While contaminated water from the impoundment appears to have moved a substantial distance from the impoundment, recent evidence suggests that the upper zone of saturation is receding, particularly during periods when there is no ponded water on the northern end of the basin. Because construction of the final surface will prevent additional ponding, it is believed that the water which currently exists in the wells along the southern and western edges of the impoundment will continue to dissipate. Therefore, it is anticipated that the tailings impoundment will have no negative impacts on groundwater resources of south Texas.

CLOSURE

As of this writing, initiation of reclamation activities is awaiting approval of an application for a license amendment to permit construction. It is anticipated that the amendment will be received in time to initiate construction activities by about December 1, 1986. Completion of construction is scheduled for early February, 1987, allowing vegetation to be established prior to the fall rainy season. Water levels in the perimeter wells will be monitored during and after construction to determine if the water levels continue to decline. It is anticipated that groundwater monitoring and settlement monitoring will continue for a few years after completion of reclamation after which the facility will be turned over to the state of Texas.

INDEX